Introduction to Digital Signal Processing: A Computer Laboratory Textbook

Georgia Tech

DIGITAL SIGNAL PROCESSING
LABORATORY SERIES

Introduction to Digital Signal Processing: A Computer Laboratory Textbook

Mark J. T. Smith
Russell M. Mersereau
The Georgia Institute of Technology

John Wiley & Sons, Inc.
New York • Chichester • Brisbane • Toronto • Singapore

Acquisitions Editor	Steven Elliot
Production Manager	Joe Ford
Production Supervisor	Charlotte Hyland
Cover Design	Bonnie Cabot
Manufacturing Manager	Lorraine Fumoso
Copy Editing Manager	Deborah Herbert

Library of Congress Cataloging in Publication Data:

Smith, Mark J. T.
    Introduction to digital signal processing : a computer laboratory
textbook / Mark J. T. Smith, Russell M. Mersereau.
        p.   cm.
    Includes index.
    ISBN 0-471-51693-7 (pbk.)
    1. Signal processing—Digital techniques—Data processing.
I. Mersereau, Russell M.   II. Title.
TK5102.5.S575 1992
621.382'2—dc20   91-36466
                         CIP

Printed in the United States of America

10 9 8 7 6 5 4 3 2 1

Printed and bound by Malloy Lithographing, Inc.

Foreword

After spending decades in the research laboratory, digital signal processing (DSP) is now emerging to make a significant impact on many areas of technology. As a result, DSP is becoming a basic subject in the electrical engineering curriculum. Although numerous textbooks and reference books are available to present the theory and applications of DSP, few of these books provide much in the way of "hands-on" experience that can help a student translate equations and algorithms into insight.

Experience during the past fifteen years at the Georgia Institute of Technology in using computers with both basic and advanced courses in DSP has shown that the personal computer can be an extremely effective learning aid when it is combined with well-designed exercises and effective software support. The Georgia Tech Digital Signal Processing Laboratory Series builds on this teaching experience to provide a set of computer laboratory books that can be used either to supplement traditional classroom/textbook presentations of the subject or as a self-study aid.

The value of computer-based laboratory experience is clear. However, just what this experience should be is somewhat dependent on the computer resources available and on the computer skills of the students. The following three approaches have proved to be effective:

1. Provide the student with a program or set of programs that perform specific DSP functions. In this situation, exercises are necessarily limited to running the programs on test data and observing the results.
2. Provide the student with a set of exercises that can be carried out by using a set of macros or low-level functions that can be strung together in some type of convenient software environment. This approach has the virtue of flexibility and is much less restrictive.
3. Provide the student with test data and suggestions for projects to be carried out with whatever programming resources are available. Clearly this is the

least restrictive approach, but is the most demanding of the student's programming/computer skills.

The first approach is likely to be frustratingly limited for students who are learning fundamental concepts, but it is very appropriate when the goal is to demonstrate complex algorithms that would require a great deal of time if students were to implement them on their own. For example, digital speech processing systems often combine many basic DSP functions, and often have many parameters whose effects can only be illustrated and studied by using an elaborate program. Another example is filter design, where students can learn the properties of different approximation methods by simply applying those methods to the same set of specifications. At the opposite extreme is the third approach, which is obviously most suitable for advanced courses or independent study where appropriate computer programming skills can be required. The second approach is perhaps the best compromise for developing insight into the fundamental algorithms and concepts of DSP. Here, the emphasis should be on the design of effective exercises and, if software is provided, it should be effective, easy to use, and bug-free.

The first book in the Georgia Tech Digital Signal Processing Laboratory Series, by Smith and Mersereau, mainly takes the second approach. It provides more than 125 exercises that can be carried out under the MS-DOS operating system using software provided with the book. This software has a wide range of basic operations and functions that can be put together to perform more complex functions by using the macro capability of DOS. Alternatively, other software packages that provide comparable capabilities may be used to do the exercises with equal effectiveness. At the Georgia Institute of Technology, student response to both the software and the exercises has been extremely favorable. Students appreciate the ease with which they can begin to actually do something with what they are learning in the classroom.

There is no doubt that DSP education is moving toward the greater use of computers. Indeed, few subjects in the electrical engineering curriculum are so well suited to the use of computers in instruction. The Georgia Tech Digital Signal Processing Laboratory Series, whose authors have many years experience in teaching and research in the DSP field, is a valuable contribution to this emerging trend in electrical engineering education.

Ronald W. Schafer
Metz, France
August 1991

Preface

There are many digital signal processing textbooks that present a good discussion of the theory, provide a variety of illustrative examples, and include a wide selection of homework problems related to the major topics. The intent of this book is somewhat different. It addresses a component of the educational process that is generally excluded in introductory books—exposure to digital signal processing in a computer environment. Computer-based exercises have been a very important component in the digital signal processing course offerings at the Georgia Institute of Technology and strongly contribute to an enriched understanding of the material. These courses are taught at the senior and first year graduate levels. This book is dedicated to providing exposure to digital signal processing in a computer environment and is intended to help the reader develop confidence in manipulating discrete-time signals. It can be used to complement a digital signal processing (DSP) textbook, as the textbook for an introductory *laboratory* course in digital signal processing, or as a self-paced introduction to DSP basics.

The introductory material typically covered in a two-course sequence on this topic is divided between two laboratory books in the Georgia Tech series: this one and *Digital Filters: A Computer Laboratory Text*. The organization of topics is designed to complement many popular introductory texts in digital signal processing. A topical reference chart is included following this preface that shows how the chapters in this book and its companion text would best complement those in several commonly used DSP textbooks.

This book includes a summary of many of the concepts basic to signal processing, a series of projects and exercises to reinforce these concepts, and a library of DSP computer functions that run on personal computers using the MS-DOS operating system. The exercises and projects integrate the computer into the learning environment and provide a level of practical experience. Although writing programs can be very helpful in understanding the intricacies of digital processing, it is also time-consuming and often results in an inefficient use of study time. The philosophy underlying this text is to provide the DSP newcomer with the experience of manipulating discrete signals without having to write and debug large programs. In fact, with few exceptions, the DSP functions included with the text enable the reader to avoid having to write programs altogether.

The software on the enclosed disk contains two executable programs: **x.exe**, which performs a diverse set of elementary signal processing functions, and **helpx.exe**, which provides an on-line description of the functions. The **helpx.exe** program may be executed by typing **helpx**. DSP functions can be easily linked together in macros (or batch files) to simulate a wide variety of systems. Although the software will generally run on any personal computer that supports the MS-DOS operating system, it has been primarily tested on IBM PS/2 and HP Vectra PC systems. It is strongly recommended that the computer supporting the DSP software have a floating-point coprocessor. The programs will also run more efficiently if all of the software and data files reside on a hard disk. The graphics functions should operate properly on most EGA-, CGA-, and VGA-equipped machines.

Organization of the Text

Each chapter begins with a brief summary of the fundamentals of the topic. These discussions are followed by a set of illustrative exercises that provide a mix of theoretical, experimental, and occasionally, programming problems. Many of the problems are straightforward, and their solutions can be easily and quickly verified by using the computer. The exercises also include a number of more difficult problems to challenge the learner. Certain exercises can be selectively omitted without a loss of understanding, according to the reader's level of familiarity with and interest in a particular topic.

The text is organized so that proceeding through the initial chapters and exercises in order provides a smooth introduction to the software that is used throughout the text. Chapters can be selected out of order, but it is suggested that Chapter 1 be reviewed first. It provides a number of useful examples that illustrate how to generate signals, how to represent signals for display and processing, and how to write macros for implementing complex systems. The last section of Chapter 1 provides a description of the DSP functions that are provided in the software environment. This same documentation is also provided in the form of an on-line help function.

The introduction to digital signal processing begins in Chapter 2 with an examination of signals and systems. This chapter also introduces the computer functions by leading the reader through a series of very basic exercises. Many properties of signals and systems, including linearity, stability, and time invariance, are examined. The exercises explore the manipulation of signals in their various representations and present the concepts of convolution and difference equations.

Chapter 3 introduces the discrete-time Fourier transform. Signals and systems are then examined in both the time and frequency domains. Chapter 4 discusses sampling and the discrete-time processing of continuous-time signals. The z-transform is introduced in Chapter 5. Representations of systems in terms of flow graphs, difference equations, poles and zeros, the discrete-time Fourier transform, and the z-transform are emphasized. Chapter 6 introduces the discrete Fourier transform or DFT. Many exercises are provided to use and interpret the DFT. The fast Fourier transform (FFT) algorithms for evaluating the DFT are also introduced in this chapter. The final chapter of the book is a collection of projects. These use the concepts and techniques discussed throughout the text to solve specific problems.

In the development of this laboratory text we were fortunate to have received many valuable contributions from colleagues and students. We gratefully acknowledge Dr. Demetrius Paris who, in his service as School Director, provided us with the support to begin this project. The software has undergone much revision and modification during the course of development. We are indebted to Dr. Steven L. Eddins who, as a graduate student, was instrumental in the development of the core computer programs and to Mr. Faouzi Kossentini, who over the last two years revised, maintained, and expanded the software as the book developed. We thank Mr. Soon-joo Hwang and Mr. Tsai Chi Huang for their contributions to the software graphics. We are also grateful for suggestions from Dr. Stanley Reeves and for the helpful comments of Paula and Dimitri Stein, Sabastian Digrande, Stanley Snow, Richard Johnson, and Gerand McDowell. Finally we thank our wives Cynthia Y. Smith and Martha Mersereau, and our children Stephen, Kevin, and Jennifer Smith and Adam and David Mersereau for their steadfast love and support over the years.

Mark J. T. Smith
Russell M. Mersereau

Reference Chart to Signal Processing Books

This book and its companion on digital filters were written to be laboratory texts and not to be the primary text in a lecture course. Several current primary texts are listed below. These can be relied on for more complete discussions, examples, and derivations of the key results that we have only summarized. The tables on the following page indicate which chapters in the primary texts provide overlapping coverage with the chapters in these two laboratory texts.

[1] L. B. Jackson, *Digital Filters and Signal Processing* (2nd), Kluwer Academic Publishers, Boston, 1989.

[2] R. Kuc, *Introduction to Digital Signal Processing*, McGraw-Hill, New York, 1988.

[3] L. C. Ludeman, *Fundamentals of Digital Signal Processing,* Harper & Row, New York, 1986.

[4] A. V. Oppenheim and R. W. Schafer, *Discrete-Time Signal Processing*, Prentice-Hall, Englewood Cliffs, NJ, 1989.

[5] A. V. Oppenheim and R. W. Schafer, *Digital Signal Processing*, Prentice-Hall, Englewood Cliffs, NJ, 1975.

[6] J. G. Proakis and D. G. Manolakis, *Introduction to Digital Signal Processing*, Macmillan, New York, 1988.

[7] R. A. Roberts and C. T. Mullis, *Digital Signal Processing*, Addison-Wesley, Reading, MA, 1987.

[8] R. D. Strum and D. E. Kirk, *First Principles of Discrete Systems and Digital Signal Processing*, Addison-Wesley, Reading, MA, 1988.

M.J.T.S.
R.M.M.

This book	Ch.1	Ch.2	Ch.3	Ch.4	Ch.5	Ch.6
Jackson	——	Ch.2	Ch.4	Ch.6	Ch.3	Ch.7
Kuc	——	Ch.2	Ch.3	Ch.3	Ch.5	Ch.4
Ludeman	——	Ch.1	Ch.1	Ch.1	Ch.2	Ch.6
Opp. & Sch., 1989	——	Ch.2	Ch.2,5	Ch.3	Ch.4	Ch.8,9
Opp. & Sch., 1975	——	Ch.1	Ch.1	Ch.1	Ch.2,4	Ch.3,6
Proakis & Manolakis	——	Ch.2	Ch.4	Ch.1	Ch.3	Ch.9
Roberts & Mullis	——	Ch.2	Ch.4	Ch.4	Ch.3	Ch.4,5
Strum & Kirk	——	Ch.3	Ch.4	Ch.2	Ch.5,6	Ch.7,8

Digital Filters	Ch. 1	Ch. 2	Ch. 3	Ch. 4	Ch. 5
Jackson	——	Ch. 9	Ch. 8	Ch. 13	Ch. 5, 11
Kuc	——	Ch. 9	Ch. 8	——	Ch. 6, 10
Ludeman	——	Ch. 3	Ch. 3,4	——	Ch. 6, 10
Oppenheim & Schafer, 1989	——	Ch. 7	Ch. 7	Ch. 10	Ch. 6
Oppenheim & Schafer, 1975	——	Ch. 5	Ch. 5	——	Ch. 4,8
Proakis & Manolakis	——	Ch. 8	Ch. 8	——	Ch. 10, 7
Roberts & Mullis	——	Ch. 6	Ch. 6	——	Ch. 9, 10
Strum & Kirk	——	Ch. 9	Ch. 10	——	Ch. 11

Contents

Introduction 1

1.1 GETTING STARTED

In this text, signal processing operations are presented in a hands-on personal computer environment in which signals are created and displayed with only a few commands. To get started you will need the following:

- An IBM-compatible personal computer running the Microsoft DOS (MS-DOS) operating system. A computer with at least an 80286 or 80386 microprocessor is preferred, although not necessary. It is also recommended that the computer contain the 80287 or 80387 floating-point coprocessor, which will dramatically increase the speed of the software.

- A hard disk. Since files will routinely be created in doing the exercises, it is suggested that a few megabytes of disk be available in the working directory.

- The MS-DOS operating system.

- Either CGA, EGA, or VGA display capability.

- The computer should contain a minimum of 640 kbytes of RAM memory. Since the digital signal processing (DSP) software will utilize a sizable part of this memory, avoid running other memory resident programs during your work session with the software. The presence of these programs in RAM memory will reduce the amount of memory available for DSP software use and thus may cause errors.

A standard ASCII text editor is also useful. It will allow you to write macros and to edit DSP files. On a small number of occasions, you are asked to write programs. This will require having a compiler for a programming language of your choice available. A *print-screen* program that allows for hard copies might also be useful in some cases, but is not necessary. The DSP software does not provide the capability to print graphics outputs. You will generally be asked to draw sketches of displayed signals from the screen.

Now create a DSP directory on your hard disk and copy all of the files on the enclosed floppy disk into that directory. It is advisable to keep an extra backup copy of the software in the event that necessary files are accidentally deleted.

The program **x.exe** contains a set of basic DSP functions that allow signals to be created and manipulated easily. For example, you can do simple operations such as adding, subtracting, or multiplying signals, but you can also perform more complex operations involving Fourier transforms or filtering. Each of these operations or functions can be invoked by simply typing **x** followed by the function name, e.g., **x add, x subtract, x multiply,** etc. In addition, these basic DSP operations can be used to design a wide variety of complex systems by creating macros composed of concatenated DSP functions. A list of the basic functions for this book is provided below.

x add	x cartesian	x cconvolve	x cexp
x conjugate	x convert	x convolve	x counter
x cshift	x dft	x divide	x dnsample
x dtft	x endloop	x extract	x fdesign
x fft	x filter	x gain	x histogram
x idft	x ifft	x imagpart	x lccde
x log	x look	x look2	x lshift
x mag	x multiply	x nlinear	x phase
x polar	x polezero	x qcyclic	x quantize
x realpart	x reverse	x revert	x rgen
x rooter	x rootmult	x siggen	x slook2
x snr	x subtract	x summer	x sview2
x truncate	x upsample	x view	x view2
x zeropad			

There are several special functions that are only intended for use in the Chapter 4 exercises. These are **x aconvolve, x afilter, x alook, x ft, x sample,** and **x sti**. The set of DSP functions provided is by no means intended to be complete. You can always add to these functions (by creating macros) to further increase the capabilities of the software whenever desired.

Before going any further, it is appropriate to explain how the DSP functions are intended to be used. Throughout this text you will be creating and processing signals. Signals may be created by typing "**x siggen**" followed by a carriage return. Try it. A menu will appear that provides options for creating a variety of different signals. For example, to create a 64-point segment of the sine wave, $x[n]$, given by

$$x[n] = \sin \frac{2\pi}{20} n, \qquad 0 \le n \le 63,$$

you would first select option **3** (sine wave) on the **x siggen** menu. Then the program will ask for the amplitude, the frequency "alpha" (in radians), the phase "phi" (also in radians), the length of the signal, and the starting point. For this example you should type **1** for the amplitude, **.314** for the frequency "alpha" (which is the approximate numerical equivalent of $\frac{2\pi}{20}$), **0** for the phase "phi," **64** for the length, and **0** for the starting point. Signals are stored in files. Therefore, **x siggen** will also ask you for the name of a file in which to store $x[n]$. To avoid confusion it is recommended that you use a simple file-naming strategy. For example, you might choose **xn** as a file name for $x[n]$.

After creating a signal, you may wish to look at it. This can be done using the function **x view**. To display the signal $x[n]$ (which you stored in file **xn**), you would type

x view   xn

and then press the "*enter*" key. To exit this function, press either the "*esc*" or "*q*" key.

To invoke any of the DSP functions, just type the function name. You will then be asked for arguments such as the input file, the output file, and function parameters if appropriate. Alternatively, these can be included on the command line as shown below:

x function   arg1   arg2 $\cdots$.

The ordering of the arguments will vary from function to function, but normally the input file(s) are listed first, followed by the output file(s), and then any floating-point or integer parameters that are required. The program will ask for any arguments that you omit. The function **x view** requires only a single input file. A more typical example is **x add**. This function has two inputs and one output. The operation

$$y[n] = x[n] + h[n]$$

can be implemented by typing

x add   xn   hn   yn

if **xn** and **hn** are files containing the signals $x[n]$ and $h[n]$. The output signal, $y[n]$, will be stored in **yn** and can be displayed using **x view**. Similarly,

$$v[n] = x[n]h[n]$$

is implemented by typing

$$\textbf{x multiply} \qquad \textbf{xn} \qquad \textbf{hn} \qquad \textbf{vn}$$

Sequences can be complex valued. Complex numbers are stored as ordered pairs containing the real and imaginary parts. For example, the complex number $2.5+j5.3$ is represented by the pair 2.5 5.3. Notice that the real and imaginary parts are separated by a space. Manipulating complex sequences is similar to manipulating real ones. For example, the operation

$$y[n] = (2 + j2)v[n]$$

can be realized using the **x gain** function. Type

$$\textbf{x gain} \qquad \textbf{vn} \qquad \textbf{yn}$$

You will be prompted for the value of the gain. Specify **2   2**, which corresponds to the complex number $2 + j2$. Alternatively, the value of the gain can be specified on the command line by typing

$$\textbf{x gain} \qquad \textbf{vn} \qquad \textbf{yn} \qquad \textbf{2} \qquad \textbf{2}$$

Short descriptions of the DSP functions and their arguments are provided in the ·last section of this chapter. In addition, the enclosed software includes a help feature that provides on-line documentation for each function. This is a very useful aspect of the software, since it allows the newcomer to use the software without having to open the book. To illustrate this feature, try typing **helpx**. A list of available DSP functions will appear on the screen. If you type **helpx** followed by one of the function names, a brief description of the operation and calling arguments will be displayed. This provides a very convenient way to become familiar with the software environment and to work the exercises in this book. The next sections of this chapter will be helpful in working several exercises in later chapters. They discuss writing macros and programs that will be compatible with the software.

1.2 DSP FILE STRUCTURE

All of the signals used in the software environment are contained in files called DSP files. You can examine them by typing them to the screen using the DOS *type* command or by using the **x view** function to display them graphically. At the beginning of each DSP file are five entries that constitute the file header. These entries contain the necessary information for the DSP functions to operate properly on the signals. The header specifies such information as the length of the signal and whether the sequence values are real or complex. In most cases, the user does not need to worry about file headers, since the DSP functions automatically generate them for all of their output files. Furthermore, if you create your own signal file using the *create file* option in **x siggen**, the header is generated automatically from the data you are asked to provide. However, if you create a signal file from a text editor, it must be in the proper format and include the file header.

This text uses two classes of signals: (1) finite length sequences; and (2) infinite length sequences whose z-transforms are rational [infinite-impulse-response (IIR) filters]. Finite length sequences are composed of strings of numbers and have the form

$$b_0, b_1, b_2, \ldots, b_P, \tag{1.1}$$

while IIR filters are composed of a ratio of polynomials in the variable z and have the form

$$H(z) = z^{-S} \frac{b_0 + b_1 z^{-1} + b_2 z^{-2} + \cdots + b_P z^{-P}}{1 + a_1 z^{-1} + a_2 z^{-2} + \cdots + a_Q z^{-Q}}. \tag{1.2}$$

The finite length sequence can be thought of as a special case of the IIR filter where the denominator polynomial is a constant of value one. The file header and format accommodate both of these representations in a simple way. The first five entries of the file, which constitute the file header, are

N	length
P	numerator order
Q	denominator order
T	1–real or 2–complex
S	starting point

where the parameters N, P, Q, T, and S are integers and have the following meanings:

N is the total number of coefficient entries in the file (excluding the header).

P is the order of the numerator polynomial. This value is only used for representing infinite-impulse-response (IIR) filters. In all other cases $P = N - 1$.

Q is the order of the denominator polynomial. This value is also only used for representing IIR filters. In all other cases $Q = 0$.

T is the coefficient type. $T = 1$ indicates real coefficients and $T = 2$ indicates complex coefficients.

S is the starting point (or index) of the first sample of the signal.

The file structure is straightforward and can be illustrated by the two generic cases shown in Table 1.1. The files shown on the left and right are IIR filter and finite length sequence files, respectively, and depict the generic signals in (1.2) and (1.1). Several points are noteworthy here. First, the labels "NUMERATOR" and "DENOMINA-

TOR" appear in the files to separate coefficients corresponding to the numerator and denominator of equation (1.2). In the case of the finite sequence (1.1), the label "SEQUENCE" appears to denote the beginning of the sequence values. Second, T is most often 1, indicating real coefficients. When $T = 2$, the coefficients $b_0, b_1, \ldots, a_1, \ldots, a_Q$ are complex. Thus each coefficient is a pair of values, the first corresponding to the real part and second to the imaginary part. A space (not a comma) is used to separate the real and imaginary parts. Third, the leading coefficient of the denominator polynomial is always assumed to be unity as shown in equation (1.2). Since the leading coefficient is always unity, it is not included in the file. The first term of the denominator is a_1. Finally, a signal file can always be examined by using the MS-DOS *type* command to display it on the screen. The readable format of the file, which includes the descriptive labels in the file header, allows the parameters and all the signal information to be easily identified by inspection. In cases where you are generating the file header with a text editor or are writing a program that creates files, it is useful to know that the descriptive labels are not case sensitive. Labels may be either uppercase or lowercase. In fact, they do not even have to be spelled correctly in order for the DSP software to recognize the files. What is important is that spaces be placed between numbers and descriptive labels, and that the number of words on each line is the same as is shown in the examples.

Table 1.1. File Structures for IIR Filters and Sequences

N	length	N	length
P	numerator order	$N - 1$	numerator order
Q	denominator order	0	denominator order
T	1–real or 2–complex	T	1–real or 2–complex
S	starting point	S	starting point
NUMERATOR		SEQUENCE	
b_0		b_0	
$\vdots$		b_1	
		$\vdots$	
b_P		b_P	
DENOMINATOR			
a_1			
$\vdots$			
a_Q			

The length of a sequence or IIR file is limited by the random-access memory available on your computer. Generally, most functions can handle sequence lengths on the order of a few thousand samples.

As a simple example of file format, try writing the DSP file for the IIR filter

$$H(z) = \frac{1 + 2z^{-1}}{1 - 0.5z^{-1}}.$$

From equation (1.2) observe that $P = 1$, $Q = 1$ and $S = 0$. The coefficients of $H(z)$ are real, thus $T = 1$. Therefore, the DSP file is

```
3       length
1       numerator order
1       denominator order
1       1–real or 2–complex
0       starting point
NUMERATOR
1
2
DENOMINATOR
−0.5
```

1.3 WRITING MACROS

Macros are batch files and have a **.bat** extension. They execute a series of commands in succession and can be used to combine DSP functions to perform more complex operations. For example, consider the system defined by the equation

$$y[n] = x^2[n] + 3.8|x[n]|$$

where $x[n]$ is the input and $y[n]$ is the output. There exists no simple DSP function to perform this operation, but it can be implemented using a short sequence of commands. If this operation is to be performed many times it is appropriate to create a macro to execute these commands automatically. This can be done by creating a file with a **.bat** extension that contains a list of the required DSP functions in their proper order. The macro can be executed by simply typing its name. It will then sequentially execute its DSP functions.

Input and output files as well as parameters can be specified on the command line of the macro. The first command line file or parameter should be specified as **%1**, the second as **%2**, the third as **%3**, and so on. The files and parameters that are typed as part of the macro command line will automatically replace these variables when the macro is executed. To illustrate all of this, let the macro **system.bat** implement the operation in our example. It has one input and one output and can be executed by typing

system   xn   yn

where **xn** and **yn** are files for the input and output, respectively. The macro consists of the following lines:

```
x multiply   %1   %1   junk1
x mag   %1   junk2
x gain   junk2   junk2   3.8
x add   junk2   junk1   %2
del junk1
del junk2
```

When the macro is executed, the file **xn** is substituted for **%1** and **yn** is substituted for **%2**. Thus the first line of the macro multiplies the input with itself to produce $x^2[n]$ and stores the result in the file **junk1**. The next line computes the magnitude of $x[n]$ and stores the result in **junk2**. The **x gain** function multiplies $|x[n]|$ by the constant 3.8 and puts the result back into file **junk2**. The signals in **junk1** and **junk2** are then added together and stored in **yn**. To avoid accumulating extraneous files, the temporary files **junk1** and **junk2** are deleted as a final step.

Macros may also be included inside of macros. For example, if you had a macro called **scale.bat** consisting of

```
x mag   %1   junk3
x gain   junk3   %2   3.8
del junk3
```

that performs the magnitude and gain operations (i.e., $3.8|x[n]|$), it could be substituted for the second and third line in the macro **system.bat**. The resulting macro would be

```
x multiply   %1   %1   junk1
command /c   scale   %1   junk2
x add   junk2   junk1   %2
del junk1
del junk2
```

MS-DOS will allow internal macros to be executed if they are preceded by the command **command /c**. For MS-DOS versions 3.3 and higher **command /c** may be replaced by **call**.

There are several useful subcommands that can be used in macros. One of these is the **echo** command. This command followed by a message will print that message to the screen during execution of the macro. The commands **echo on** and **echo off** will, respectively, enable and disable the display of commands during the execution of a macro. For example, the inclusion of **echo off** at the beginning of your macro will prevent commands from appearing on the screen. However, all messages from programs called within the macro will be displayed.

A colon "**:**" inserted at the beginning of a line can be used to disable any command that follows it on that line. This provides a convenient way to work with macros that are continually being modified. For a more complete discussion on the topic of writing macros, consult the section on batch file commands in your MS-DOS manual. There may be some variation in these commands, depending on the version of DOS being used.

Digital algorithms often involve successive invocations of a particular operation. This is called *iteration*. Scientific programming languages accomplish this by allowing operations to be placed in loops. MS-DOS does not provide an elegant way to perform repetitive operations. One approach is to manually invoke a macro that represents one iteration of an operation. Another is to embed a key macro in a shell that invokes it several times. But these approaches are cumbersome and do not conveniently lend themselves to changing the number of iterations. To aid in implementing iterative operations the software contains the two operators **x counter** and **x endloop**. The use of these two operators is indicated by the macro below.

```
ECHO OFF
x counter 25
:LOOP
(DSP functions)
x endloop
IF EXIST COUNT.### GOTO :LOOP
ECHO ON
```

The DSP functions shown on the fourth line of the macro represent the desired string of DSP functions that implement one iteration of the algorithm. The line **x counter 25** creates a file called COUNT.### that contains the number 25 (the number of iterations to be performed). Alternatively, the number of iterations may be given on the command line, in this case by using **%1** in place of **25** in the macro and specifying **25** on the command line.

The **x endloop** command opens the file COUNT.###, subtracts 1 from it, and, if the result is not zero, writes the new value back into the file. If the count has been decremented to zero, the file COUNT.### is removed. The "IF" statement in line 6 simply looks to see whether or not the file exists.

EXERCISE 1.3.1. **Practice with Macros**

Macros permit you to create your own useful functions by concatenating DSP functions. As an illustration, consider the operation of taking the complex conjugate. The complex conjugate of a sequence

$$x[n] = a[n] + jb[n]$$

is $x^*[n]$ and is defined to be

$$x^*[n] = a[n] - jb[n].$$

This operation is an explicit function in the DSP software called **x conjugate**, but it can also be realized by using several other DSP functions in series. Write a macro that will accept a complex input sequence as a file and produce an output file that is its complex conjugate.

A solution to this exercise is given at the very end of this chapter.

EXERCISE 1.3.2. **Practice in Writing Signal Files**

On several occasions you may wish to write your own computer program (compatible with the DSP software) for processing files. In such a case, it is important to be familiar with the file structure. Here some simple signals are presented for practice in writing DSP files.

(a) Using a text editor, write a signal file for the sequence 0, 1, 2, 3, 4, 5. Assume that the sequence starts at zero. Examine the file using **x view** to verify that it is correct and that it can be read properly by the software. Now check your answer by using the *ramp* option in **x siggen**. You may check your answer by typing this file to the screen and comparing.

(b) Determine the signal file header for the IIR filter

$$H(z) = \frac{1 + z^{-1} + z^{-2} + z^{-3} + z^{-4} + z^{-5} + z^{-6}}{1 - z^{-1} - z^{-2} - z^{-3} - z^{-4} - z^{-5} - z^{-6}}.$$

Note that you are only being asked for the file header (i.e., the first five entries of the file). To check your answer, type the file **hh** (which is provided for you on the disk) to the screen. This file has the same file header as $H(z)$.

Answers may also be checked by using the *create file* option in **x siggen**.

1.4 DSP FUNCTION DOCUMENTATION

Each of the DSP functions can be invoked by simply typing its name. The program will ask you for any input files, output files, and parameters that may be required. This information can also be provided on the command line. This is particularly important for writing macros. This section provides descriptions of the DSP functions used for this text and describes all of their command-line arguments and parameters. Much of this same information may be accessed on line using the **helpx** function. To use this feature of the software, simply type **helpx**.

The functions operate on both real and complex sequences unless explicitly stated otherwise in the documentation. The functions are designed to process sequences. Several of the functions also handle IIR filters. These are **x convert, x dtft, x filter, x polezero**, and **x revert**. IIR filters may be processed by using **x**

convert to put the numerator and denominator polynomials into separate files that can be processed individually as sequences.

A file can generally have any name. However, several special files are included for your use in specific exercises. These files are **aa, bb, cc, ccc, dd, ee, ff, gg, hh**, and **sig1**. Care should be taken to avoid overwriting these files. In addition, the macros **system1.bat**, **system2.bat**, and **system3.bat** are predefined systems provided for your use. Thus you should avoid using these names.

x add

The **x add** function adds two sequences together in a point-by-point fashion. In other words, it implements the equation

$$y[n] = x_1[n] + x_2[n]$$

and is executed by typing

$$\mathbf{x\ add}\quad \overbrace{(\text{input1})}^{x_1[n]}\quad \overbrace{(\text{input2})}^{x_2[n]}\quad \overbrace{(\text{output})}^{y[n]}$$

x cartesian

The **x cartesian** function is used to convert a complex sequence from polar to Cartesian form. It is executed by typing

$$\mathbf{x\ cartesian}\quad \overbrace{(\text{input})}^{x[n]}\quad \overbrace{(\text{output})}^{y[n]}$$

The complex finite length signal $x[n]$ is stored in polar form, in which the magnitude of each sample of $x[n]$ is stored in the real part and the phase is stored in the imaginary part of each entry. The output sequence is the same sequence stored in Cartesian (real–imaginary) form. More precisely, **x cartesian** implements the relationship

$$y[n] = \overbrace{|x[n]|\cos(\angle x[n])}^{\Re e\{y[n]\}} + j\overbrace{|x[n]|\sin(\angle x[n])}^{\Im m\{y[n]\}}.$$

The functions **x polar** and **x cartesian** are inverses of each other.

x cconvolve

The **x cconvolve** operation is executed by typing

$$\overset{x[n]}{\mathbf{x\ cconvolve}}\quad \overbrace{(\text{input1})}^{x[n]}\quad \overbrace{(\text{input2})}^{h[n]}\quad \overbrace{(\text{output})}^{y[n]}\quad N$$

This function implements the N-point circular convolution of two finite length sequences:

$$y[n] = \sum_{m=0}^{N-1} x[m]h[(n-m) \bmod N] \quad n = 0, 1, \ldots, N-1.$$

The value of N must be an integer that is greater than or equal to the length of the longer sequence. Both inputs are assumed to start at zero.

x cexp

The **x cexp** function multiplies the finite length input sequence $x[n]$ by a complex exponential. It is executed by typing

$$\mathbf{x\ cexp}\quad \overbrace{(\text{input})}^{x[n]}\quad \overbrace{(\text{output})}^{y[n]}\quad \omega_o$$

The output sequence $y[n]$ is defined by

$$y[n] = x[n]e^{j\omega_0 n}$$

where ω_0 is the frequency in radians; ω_0 can be complex.

x conjugate

The **x conjugate** function takes the complex conjugate of a sequence. It is executed by typing

$$\mathbf{x\ conjugate}\quad \overbrace{(\text{input})}^{x[n]}\quad \overbrace{(\text{output})}^{y[n]}$$

where $y[n]$ is the complex conjugate of $x[n]$.

x convert

The **x convert** function can be useful for manipulating IIR filters. It is executed by typing

$$\mathbf{x\ convert}\quad \overbrace{(\text{input})}^{H(z)}\quad \overbrace{(\text{output1})}^{B(z)}\quad \overbrace{(\text{output2})}^{A(z)}$$

It accepts, in its input file, the coefficients of an IIR filter in the form $H(z) = B(z)/A(z)$ and returns the numerator polynomial $B(z)$ and denominator polynomial $A(z)$ in separate files. These can then be processed separately. The numerator and denominator files can be recombined using the function **x revert**.

x convolve

The **x convolve** function forms the linear convolution of two sequences. It is executed by typing

$$\textbf{x convolve} \quad \overbrace{(\text{input1})}^{x[n]} \quad \overbrace{(\text{input2})}^{h[n]} \quad \overbrace{(\text{output})}^{y[n]}$$

The convolution of the two sequences $x[n]$ and $h[n]$ is defined as

$$y[n] = x[n] * h[n] = \sum_{m=-\infty}^{\infty} x[m]h[n-m].$$

The **x convolve** function performs the convolution of two finite length input sequences and stores the result in the output file. Since convolution is a commutative operation, the order in which the input files are given does not matter.

x counter

The **x counter** function is used with the **x endloop** function for implementing iterative procedures and recursions. It is called by the statement

$$\textbf{x counter} \quad N$$

It will create a file called COUNT.### and store the value of N in it. The value of N should be a positive integer. Additional information on implementing iterative procedures is given in Section 1.3.

x cshift

The **x cshift** function is executed by typing

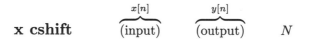

$$\textbf{x cshift} \quad \overbrace{(\text{input})}^{x[n]} \quad \overbrace{(\text{output})}^{y[n]} \quad N$$

This function performs an N-point circular right shift on the input sequence. The value of N is the number of points shifted. When N is negative, the shift becomes a circular left shift. If L denotes the length of the sequence $x[n]$, the **x cshift** function implements the equation

$$y[n] = x[(n - N) \bmod L].$$

This function requires that the starting point of the input be at the origin.

x dft

The **x dft** function calculates the N-point discrete Fourier transform of its input sequence. It is executed by typing

$$\mathbf{x\ dft} \quad \overbrace{\text{(input)}}^{x[n]} \quad \overbrace{\text{(output)}}^{X[k]}$$

The function performs a direct implementation of the DFT summation

$$X[k] = \sum_{n=0}^{N-1} x[n]e^{-j(2\pi/N)nk}, \qquad 0 \le k \le N - 1.$$

The starting point of the input must be zero and N must be a positive integer. The inverse DFT can be performed using **x idft**. The value of the sequence length N, is read from the file header.

x divide

The **x divide** function performs a point-by-point division of two sequences. It is executed by typing

$$\mathbf{x\ divide} \quad \overbrace{\text{(input1)}}^{x[n]} \quad \overbrace{\text{(input2)}}^{v[n]} \quad \overbrace{\text{(output)}}^{y[n]}$$

The functional form of the operation is

$$y[n] = x[n]/v[n].$$

When a zero is encountered in $v[n]$, the function returns an error message.

x dnsample

The **x dnsample** function performs a downsampling of a sequence by an integer rate. The output sequence is defined by

$$y[n] = x[Mn]$$

where M is a positive integer. The function may be executed by typing

$$\mathbf{x\ dnsample} \quad \overbrace{\text{(input)}}^{x[n]} \quad \overbrace{\text{(output)}}^{y[n]} \quad M$$

x dtft

The **x dtft** function computes the discrete-time Fourier transform $H(e^{j\omega})$. The input can be either a finite length sequence or the coefficients of an IIR filter. The finite-impulse response (FIR) filter lengths and IIR filter orders should be less than 512 for proper operation. The output is a plot of the function

$$H(e^{j\omega}) = \sum_{n=-\infty}^{\infty} h[n]e^{j\omega n} \qquad -\pi \leq \omega \leq \pi$$

in the frequency range $\omega = -\pi$ to $\omega = \pi$. Note that the frequency origin ($\omega = 0$) appears in the middle of the screen. The function is called by the instruction

$$\textbf{x dtft} \qquad \overbrace{\text{(input)}}^{h[n]}$$

A display option menu appears that asks for display of the magnitude, log magnitude, phase, real part, imaginary part, magnitude and phase, or real and imaginary parts. Once the plot is displayed on the screen, three keys are available to impart commands. Pressing the "*g*" key allows you to turn the grid of tic marks on or off to suit your preference. The "*esc*" key can be used to return you to the display option menu. The "*q*" key will terminate the display immediately and return you to DOS. In addition, the function creates a complex signal file, _dtft_. It is an intermediate file containing 513 samples (from -256 to $+256$) that represent the discrete-time Fourier transform of the signal. This is the signal that is displayed by the graphics routine. It can be examined using a text editor to obtain more detailed information not conveniently seen from the graphics display.

x endloop

The function **x endloop** is used for programming iterative operations. It is called by a statement of the form

$$\textbf{x endloop}$$

It looks for the file COUNT.###, decrements its contents by one, and deletes the file if the result of the decrement is zero. Its use is described more fully in Section 1.3.

x extract

The **x extract** function extracts a block of samples from a longer input sequence. It can be executed by typing

$$\mathbf{x}\ \mathbf{extract}\qquad \overbrace{(\text{input})}^{x[n]}\qquad \overbrace{(\text{output})}^{y[n]}\qquad L\quad P\quad M$$

where L is the length of the block to be extracted, P is the index or starting point of the first sample to be extracted, and M is the index to which that sample is to be assigned in the output sequence, i.e., the starting point for the extracted block in the output sequence.

x fdesign

The **x fdesign** function is a menu-driven program that designs FIR lowpass and highpass filters. These filters are simply finite length sequences whose Fourier transform magnitudes are approximately unity in one region of the spectrum and zero in the remaining region. The function **x fdesign** can be executed by typing

$$\mathbf{x}\ \mathbf{fdesign}\qquad (\text{output})$$

The filter design is based on the window design procedure, and three different window options are presented: the rectangular window, the Hamming window, and the Hanning (or von Hann) window. You will be asked for the filter length and the filter cutoff frequency in radians.

A variety of more sophisticated programs for designing digital filters are available in the companion text *Digital Filters: A Computer Laboratory Text*.

x fft

The **x fft** function evaluates the discrete Fourier transform (DFT) using the Cooley-Tukey fast Fourier transform (FFT) algorithm. It can be executed by typing the command

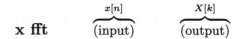

$$\mathbf{x}\ \mathbf{fft}\qquad \overbrace{(\text{input})}^{x[n]}\qquad \overbrace{(\text{output})}^{X[k]}$$

The length of the input sequence must be a positive integer power of 2. The length of the FFT is the same as the length of the input sequence, and the input sequence must always begin at $n = 0$. The inverse DFT can be evaluated efficiently using the function **x ifft**.

x filter

The **x filter** function implements either an IIR or an FIR filter. It can be run by typing the command

$$\mathbf{x\ filter}\quad \underbrace{(\text{input1})}_{x[n]}\quad \underbrace{(\text{input2})}_{h[n]\ \text{or}\ H(z)}\quad \underbrace{(\text{output})}_{y[n]}\quad L$$

The first signal $x[n]$ is the finite length input sequence. The second input contains the coefficients of the IIR filter or FIR filter. Thus the filtering operation can be viewed as

$$y[n] = x[n] * h[n].$$

Since the output of an IIR filter is normally of infinite duration, you must also specify the number of output samples to be calculated. This is the last command-line parameter, L.

The program computes L samples of the output. This function is virtually identical to **x lccde** (with zero initial conditions), except that the filter coefficients are read from a file.

x gain

The **x gain** function multiplies a finite length sequence by a user-specified real or complex constant G. It can be run by typing

$$\mathbf{x\ gain}\quad \underbrace{(\text{input})}_{x[n]}\quad \underbrace{(\text{output})}_{y[n]}\quad G$$

The input and output sequences are related by

$$y[n] = Gx[n].$$

To specify a complex gain, the real and imaginary parts of G should be entered as two real numbers separated by a space.

x histogram

The **x histogram** function computes the histogram of the input and puts it in the output file. The histogram is a function whose abscissa corresponds to amplitude values of the input and whose ordinate represents the number of occurrences of that amplitude value in the input. Histograms are typically used as a representation of the probability density function.

The function is executed by typing

$$\mathbf{x\ histogram}\quad \underbrace{(\text{input})}_{x[n]}\quad \underbrace{(\text{histogram})}_{y[n]}$$

Since the dynamic range of the input is inherently very large and would imply a histogram length of the same size, limits are placed on the amplitude range of the input. Specifically, the amplitude range of the input must be between -500 and 500. If input values occur outside of this range, an error message will result. In such a case, you may wish to scale the input prior to execution so that it fits within the allowable input range. All input values are quantized (rounded) to the nearest integer value.

x idft

The **x idft** function calculates an inverse DFT. This is the inverse function of **x dft**. It is executed by typing

$$\mathbf{x\ idft} \qquad \overbrace{(\text{input})}^{X[k]} \qquad \overbrace{(\text{output})}^{x[n]}$$

The function **x idft** performs the N-point inverse discrete Fourier transform on the input sequence by directly evaluating the sum

$$x[n] = \frac{1}{N} \sum_{k=0}^{N-1} X[k] e^{j(2\pi/N)nk}, \qquad 0 \le n \le N-1.$$

The starting point of $X[k]$ must be zero and N must be a positive integer.

x ifft

The function **x ifft** evaluates the inverse discrete Fourier transform. It is similar to **x idft** except that the computation is considerably more efficient and the length of the transform, N, is restricted to be a power of two. It can be executed by typing

$$\mathbf{x\ ifft} \qquad \overbrace{(\text{input})}^{X[k]} \qquad \overbrace{(\text{output})}^{x[n]}$$

The input sequence must always start at $k = 0$. The functions **x fft** and **x ifft** are inverses of each other.

x imagpart

The function **x imagpart** takes the imaginary part of a complex input signal $x[n]$. The output sequence is real. The calling statement is

$$\mathbf{x\ imagpart} \qquad \overbrace{(\text{input})}^{x[n]} \qquad \overbrace{(\text{output})}^{y[n]}$$

x lccde

The **x lccde** function is an interactive function that implements a linear constant coefficient difference equation of the form

$$y[n] = -a_1 y[n-1] - a_2 y[n-2] - \cdots - a_N y[n-N]$$
$$+ b_0 x[n] + b_1 x[n-1] + \cdots + b_M x[n-M].$$

It can be executed by typing

$$\textbf{x lccde} \quad \overbrace{(\text{input})}^{x[n]} \quad \overbrace{(\text{output})}^{y[n]}$$

This is a user interactive function. You are prompted for the values of N and M as well as for the individual coefficients $b_0, b_1, \ldots, b_M$ and $a_1, a_2, \ldots, a_N$. In addition, you will be asked to specify whether the coefficients are real or complex. Since the output is normally infinitely long, you will also be asked to specify the number of output values to be computed.

x log

This function computes the natural logarithm of a finite length input, i.e.,

$$y[n] = \ln x[n].$$

The input may be either real or complex. It can be executed by typing

$$\textbf{x log} \quad \overbrace{(\text{input})}^{x[n]} \quad \overbrace{(\text{output})}^{y[n]}$$

Input sample values that are zero are not permitted and will result in an error message.

x look

The **x look** function is executed by typing

$$\textbf{x look} \qquad (\text{input})$$

It plots a discrete set of points as a continuous function. If the sequence is complex, a display option menu will appear that will prompt you for display of the magnitude, log magnitude, phase, real part, imaginary part, magnitude and phase, or real and imaginary parts. Once the plot is displayed on the screen, three keys are available to impart commands. Pressing the "*g*" key allows you to turn the grid of tic marks on or off to suit your preference. The "*esc*" key will return you to the display option menu for the case of complex signals. It will simply terminate the display if the signal is real. The "q" key will terminate the display in either case and return you to DOS. This function is intended for use on CGA-, EGA-, and VGA-equipped systems.

x look2

The **x look2** function is executed by typing

$$\text{x look2} \qquad (\text{input1}) \qquad (\text{input2})$$

It is identical to the **x look** function, but allows you to display two plots at a time, one above the other.

x lshift

The **x lshift** function can be executed by typing

$$\text{x lshift} \qquad \overbrace{(\text{input})}^{x[n]} \qquad \overbrace{(\text{output})}^{y[n]} \qquad D$$

This function performs a linear right shift of the finite length input sequence $x[n]$ by D samples when D is positive, i.e.,

$$y[n] = x[n - D].$$

The shift is a left shift when D is negative. D must be an integer in the range $-32,768$ to $32,767$.

x mag

The function **x mag** takes the magnitude of the finite length complex input sequence and produces an output sequence that is real. Specifically, the function implements the equation

$$y[n] = |x[n]|.$$

If the input sequence is real, the **x mag** function is equivalent to an absolute-value operation. The **x mag** function can be executed by typing

$$\text{x mag} \qquad \overbrace{(\text{input})}^{x[n]} \qquad \overbrace{(\text{output})}^{y[n]}$$

x multiply

The **x multiply** function implements the equation

$$y[n] = x[n]w[n].$$

It can be executed by typing

$$\text{\textbf{x multiply}} \quad \overbrace{\text{(input1)}}^{x[n]} \quad \overbrace{\text{(input2)}}^{w[n]} \quad \overbrace{\text{(output)}}^{y[n]}$$

x nlinear

The **x nlinear** function is executed by typing

$$\text{\textbf{x nlinear}} \quad \overbrace{\text{(input)}}^{x[n]} \quad \overbrace{\text{(output)}}^{y[n]}$$

It is a menu-driven program that performs one of three predefined nonlinear operations on a finite length input. If $x[n]$ and $y[n]$ represent the input and output sequences, respectively, any of the following operations can be specified.

1. $y[n] = \sin(\alpha x[n] + \phi)$ ($x[n]$ must be real.) The argument of the sine is always assumed to be in radians.
2. $y[n] = c^{x[n]}$.
3. $y[n] = (x[n])^{\alpha}$.

The program prompts you for all necessary parameters. It may also be operated in a non-menu-driven mode, which allows it to be conveniently used in macros. This is achieved by specifying the parameters on the command line as indicated:

$$\text{\textbf{x nlinear}} \quad \overbrace{\text{(input)}}^{x[n]} \quad \overbrace{\text{(output)}}^{y[n]} \quad \text{option number} \quad \alpha \quad \phi$$

In cases where α and/or ϕ are not explicit parameters, these arguments are omitted from the command line.

x phase

The **x phase** function produces a sequence corresponding to the point-by-point phases of the complex input sequence. It is executed by typing

$$\text{\textbf{x phase}} \quad \overbrace{\text{(input)}}^{x[n]} \quad \overbrace{\text{(output)}}^{y[n]}$$

It computes the principal value of the phase at each point. Each phase value will lie in the range $(-\pi, \ \pi]$. The specific function implemented is

$$y[n] = \arctan(\Im m\{x[n]\}/\Re e\{x[n]\}),$$

which is followed by a correction based on the signs of the real and imaginary parts to extend the range of the function.

x polar

The **x polar** function is used to convert a complex sequence from Cartesian to polar form. It is executed by typing

$$\mathbf{x\ polar}\quad \overbrace{(\text{input})}^{x[n]}\quad \overbrace{(\text{output})}^{y[n]}$$

The finite length complex input signal $x[n]$ is assumed to be stored in Cartesian (real–imaginary) form. The output sequence is the same sequence stored in polar (magnitude–phase) form. More precisely, **x polar** implements the equation

$$y[n] = \overbrace{\sqrt{\Re e\{x[n]\}^2 + \Im m\{x[n]\}^2}}^{|x[n]|} + j\,\overbrace{\arctan(\Im m\{x[n]\}/\Re e\{x[n]\})}^{\angle x[n]}$$

(where a correction is applied to the phase function to extend its range). The magnitude of $x[n]$ is stored as the real part of $y[n]$, while the phase of $x[n]$ is stored as the imaginary part of $y[n]$. The functions **x polar** and **x cartesian** are inverses of each other.

x polezero

The **x polezero** function is a menu-driven program that allows you to display the pole/zero plot of a function or sequence as well as its corresponding magnitude response for frequencies in the range from $-\pi$ to π. The program is invoked by typing **x polezero**. A main menu will appear that will give you six general options:

1. Read a file;
2. Adjust circle size;
3. List poles and zeros;
4. Change poles and zeros;
5. Write to output file;
6. Quit.

Option 1. This option should be used first. It permits a file containing a sequence or IIR filter coefficients to be read. As a rule of thumb, sequences or functions should not contain more than twenty poles and zeros for proper operation. If some of the roots lie outside of the field of view for the plot, a message will appear at the top of the screen. As is true of most rooting algorithms, the leading coefficient of the input file must be nonzero in order for the program to function properly. In the case of IIR filters, the leading coefficient of the numerator must be nonzero.

Option 2. This option provides some limited control over the size of the unit circle on the display. The size of the circle is determined by an automatic scaling subroutine that attempts to provide a readable display. In some cases the subroutine will allow you to enlarge or reduce the circle by selecting the appropriate submenu option.

Option 3. This option allows you to list the poles and zeros as well as the numerator and denominator polynomial coefficients of your function or sequence.

Option 4. This option enables you to add or delete zeros and poles, or move existing zeros and poles. These operations are invoked through a seven-option submenu. Operations requiring the deletion or movement of a root and its conjugate will ask you to specify both the root and conjugate. Poles and zeros may be moved either as single roots or in complex conjugate pairs. The directional arrow keys control the movement of the roots. The initial increment in which the roots may be moved is 0.01. To get more rapid movement of the roots select the appropriate submenu options. As the root locations are modified, it is possible to display the new magnitude response simultaneously by pressing the *"d"* key. Pressing the *"enter"* key returns control to the main menu.

Option 5. This option allows you to write your new or modified sequence or function to a file. Thus **x polezero** may be used to design or modify linear time-invariant (LTI) systems via their pole/zero plot. The program is equipped to handle both real- and complex-valued functions and sequences.

x qcyclic

The **x qcyclic** function can be used to quantize the input sequence to a prespecified number of levels. It is executed by typing

$$\mathbf{x\ qcyclic}\ \ \overbrace{\text{(input)}}^{x[n]}\ \ \overbrace{\text{(output)}}^{y[n]}\ \ (X_{\min})\ \ (X_{\max})\ \ \text{(levels)}$$

The quantization is uniform. The user specifies two numbers corresponding to the smallest and largest quantization levels, $(X_{\min})$ and $(X_{\max})$, and the number of levels, which must be an integer. The function is virtually identical to **x quantize** except in the way that amplitude values that exceed the amplitude bounds $X_{\max}$ and $X_{\min}$ are handled. Multiples of the factor $|X_{\max} - X_{\min}|$ are either added or subtracted from the amplitude until it lies within the quantizer range. This simulates the effects of *wrap-around* or *cyclic* overflow in two's complement arithmetic.

x quantize

The **x quantize** function can be used to quantize the input sequence to a prespecified number of levels. It is executed by typing

$$\textbf{x quantize} \quad \overbrace{(\text{input})}^{x[n]} \quad \overbrace{(\text{output})}^{y[n]} \quad (X_{\min}) \quad (X_{\max}) \quad (\text{levels})$$

The user specifies two numbers corresponding to the smallest and largest quantization
levels, $(X_{\min})$ and $(X_{\max})$, and the number of levels, which must be an integer. The
function is virtually identical to **x qcyclic** except that inputs exceeding the maximum
quantizer limit are set to $X_{\max}$ and those falling below the lower bound are set to $X_{\min}$.
This function is used to simulate saturation arithmetic.

x realpart

The function **x realpart** takes the real part of a complex input signal $x[n]$. The
calling statement is

$$\textbf{x realpart} \quad \overbrace{(\text{input})}^{x[n]} \quad \overbrace{(\text{output})}^{\Re\{x[n]\}}$$

x reverse

The **x reverse** function forms the time reversal of the input sequence. The output
sequence is defined by

$$y[n] = x[-n].$$

The function can be called by a statement of the form

$$\textbf{x reverse} \quad \overbrace{(\text{input})}^{x[n]} \quad \overbrace{(\text{output})}^{y[n]}$$

x revert

The **x revert** function takes two sequences and combines them to form an IIR filter.
It can be executed by typing

$$\textbf{x revert} \quad \overbrace{(\text{input1})}^{B(z)} \quad \overbrace{(\text{input2})}^{A(z)} \quad \overbrace{(\text{output})}^{H(z)}$$

It treats $B(z)$ as coefficients of the numerator polynomial and $A(z)$ as coefficients of
the denominator polynomial of $H(z)$. The functions **x revert** and **x convert** can

be used to decompose and merge IIR filters for individual processing of the numerator and denominator polynomials. The leading coefficient a_0 of the denominator $A(z)$ is always assumed to be equal to unity. If the first coefficient of $A(z)$ is not one, **x revert** will divide $B(z)$ and $A(z)$ by a_0 so that the format matches equation (1.2).

x rgen

The **x rgen** function generates a random number sequence of prescribed length. The samples are uniformly distributed between -0.5 and 0.5. It is called by typing

$$\overbrace{\phantom{(\text{output})}}^{y[n]}$$
$$\textbf{x rgen} \qquad (\text{output}) \qquad (\text{length}) \qquad (\text{seed})$$

The seed parameter is an integer used to initialize the random number generator. Running **x rgen** twice with the same positive seed will reproduce the same output sequence. Running **x rgen** with a negative seed causes it to use the system clock to initialize the generator, resulting in a nonrepeatable output sequence.

x rooter

The function **x rooter** is used to evaluate the roots of polynomials whose coefficients appear in the input sequence. It can be called using the statement

$$\overbrace{\phantom{(\text{input})}}^{x[n]} \qquad \overbrace{\phantom{(\text{output})}}^{y[n]}$$
$$\textbf{x rooter} \qquad (\text{input}) \qquad (\text{output})$$

The input sequence is interpreted as coefficients of the polynomial in the variable z^{-1}, i.e.,

$$x[0] + x[1]z^{-1} + \cdots + x[P]z^{-P}.$$

The function **x rooter** factors the polynomial into the form

$$(1 - \alpha_1 z^{-1})(1 - \alpha_2 z^{-1}) \cdots (1 - \alpha_P z^{-1}).$$

The output is the sequence of complex roots $\alpha_1, \alpha_2, \ldots, \alpha_P$ expressed in terms of their real and imaginary parts. The input sequence is assumed to be real, and the output sequence is complex in general.

Finding the roots of high-order polynomials is generally a difficult task. The degree of difficulty is often dependent on the values of the polynomial coefficients as well as the filter order. This function can usually find the roots of polynomials of order 25 or less. As with most polynomial rooting algorithms, the leading coefficient must be nonzero.

x rootmult

The **x rootmult** function is the inverse of **x rooter**. It is executed by typing

$$\text{x rootmult} \quad \overbrace{\text{(input)}}^{x[n]} \quad \overbrace{\text{(output)}}^{y[n]}$$

The input is assumed to be a list of polynomial roots given as a sequence of complex numbers. It converts these roots to first-order factors and multiplies these to form a polynomial. This function can be used with **x rooter** to factor polynomials and reconstruct them (up to a constant gain factor) from their roots. This function may produce numerical errors when the number of roots is large. However, it can be expected to perform reliably when the polynomials to be reconstructed are of order 25 or less.

x siggen

The **x siggen** function is a signal generator. It is called without arguments using the statement

$$\text{x siggen}$$

It can be used to generate square wave, triangular wave, sine wave, exponential, block, ramp, impulse train, and chirp functions, each of which is defined over a finite interval beginning at zero. However, the program allows each signal to be shifted so that it starts at a user-specified starting point. In other words, if the signal created is $x[n]$, the effect of choosing starting point n_0 is to produce the sequence $x[n - n_0]$. In addition, the program provides the option to create an arbitrary signal file. This can be either a sequence or an IIR filter. A description of each menu option is given next.

Square Wave. A square wave of period P has the functional form

$$x[n] = \begin{cases} A, & 0 \le n \le T - 1 \\ 0, & T \le n \le P - 1. \end{cases}$$

The sequence generated, $x[n]$, will have pulse width T, period P (where $P \ge T$), M periods of the square wave in the output, and amplitude A. The function will prompt you for values of the length, period, number of periods, and amplitude. All of these are integers except for the amplitude, which is real. You will also be asked for the sequence starting point, n_0, and for the name of the output file.

Triangular Wave. The form of this sequence will vary depending on whether the period P is even or odd. If an even period is specified, the triangular wave has the form

$$x[n] = \begin{cases} \frac{2An}{P}, & 0 \le n \le \frac{P}{2} - 1 \\ x[P-n], & \frac{P}{2} \le n \le P - 1. \end{cases}$$

The total sequence will consist of M periods. When an odd period is specified the wave has the form

$$x[n] = \begin{cases} \frac{2An}{P-1}, & 0 \le n \le \frac{P-1}{2} \\ x[P-n], & \frac{P+1}{2} \le n \le P - 1. \end{cases}$$

The program prompts you for the (integer) period P, the (integer) number of periods in the sequence, and the maximum amplitude A. You will also be asked to specify the starting point, n_0, and the name of the output file.

Sine Wave. The sine wave has the form

$$x[n] = A\sin(\alpha n + \phi), \qquad 0 \le n \le N - 1$$

where the argument of the sine is in radians. You will be asked for values of A, α, ϕ, and N (which must be an integer). You will also be asked to specify the starting point, n_0, and for the name of the output file.

Exponential $K e^{\alpha n}$. This exponential has the form

$$x[n] = Ke^{\alpha n}, \qquad 0 \le n \le N - 1.$$

You will be asked for values of K (which must be real), α (which can be real or complex), and N (which must be a positive integer). You will also be asked to specify the starting point, n_0, and for the name of the output file.

Exponential $K\alpha^n$. This exponential has the form

$$x[n] = K\alpha^n, \qquad 0 \le n \le N - 1.$$

You will be asked for values of K (which must be real), α (which can be real or complex), and N (which must be a positive integer). You will also be asked to specify the starting point, n_0, and for the name of the output file.

Block. The block or pulse signal has the form

$$x[n] = A, \qquad 0 \le n \le N - 1$$

where A is the block amplitude. You will be asked for the amplitude, the block length, the starting point, n_0, and for the name of the output file. The block function can be used to approximate a step.

Ramp. The ramp has the form

$$x[n] = An, \qquad 0 \leq n \leq N - 1$$

where A is the slope and N is the sequence length. The program will ask for values of the slope A and the length N. In addition you will be asked for the starting point, n_0, and for the name of the output file.

Impulse Train. The impulse train has the form

$$x[n] = \begin{cases} A, & n/P \text{ an integer} \\ 0, & \text{otherwise} \end{cases}$$

over the range $0 \leq n \leq MP - 1$. You will first be asked for the values of the amplitude and then the period P and the number of periods M, both of which must be integers. You will also be asked for the starting point, n_0, and for the name of the output file.

Chirp. The chirp signal has the form

$$x[n] = A \sin[\alpha_1 n + \alpha_2 n^2 + \phi], \qquad 0 \leq n \leq N - 1$$

where the argument of the sine is always assumed to be in radians. You will be asked for values of the amplitude A, α_1, α_2, ϕ, and N (an integer). You will also be asked for the starting point, n_0, and for the name of the output file.

Create File. The create file option allows you to create a DSP sequence or an IIR filter. The program will prompt you for the sequence length and coefficients in the case of the sequence, and for the numerator order, denominator order, and coefficients in the case of the IIR filter. You will also be asked to specify real or complex values, a starting point for the sequence or filter, and an output file name.

x slook2

The **x slook2** function that stands for *superimposed look* is executed by typing

$$\textbf{x slook2} \qquad (\text{input1}) \qquad (\text{input2})$$

It displays two real sequences in continuous display mode superimposed on the same coordinate axes. This function is only intended for real sequences. Once the plot is displayed on the screen, three commands are available. Pressing the "*g*" key allows you to turn the grid of tic marks on or off to suit your preference. The "*esc*" key can be used to return you to the display option menu. The "*q*" key will terminate the display immediately and return you to DOS. This function is intended for use on CGA-, EGA-, and VGA-equipped systems.

x snr

The **x snr** function computes the signal-to-noise-ratio (SNR) defined as

$$\text{SNR(dB)} = 10 \log_{10} \frac{\sum\limits_{n=L}^{N} x[n]^2}{\sum\limits_{n=L}^{N} (x[n] - \hat{x}[n])^2} \tag{1.3}$$

where $x[n]$ is the original signal input and $\hat{x}[n]$ is the processed or reconstructed signal that approximates the original. It is executed by typing

x snr $x[n]$ $\hat{x}[n]$

The output of this function is a single number that is the SNR. Its units are decibels. The sequences $x[n]$ and $\hat{x}[n]$ must be defined over the same range.

x subtract

The **x subtract** function implements the operation

$$y[n] = x_1[n] - x_2[n].$$

It is called by the statement

$x_1[n]$ $\qquad$ $x_2[n]$ $\qquad$ $y[n]$

x subtract $\overbrace{(\text{input1})}$ $\overbrace{(\text{input2})}$ $\overbrace{(\text{output})}$

x summer

The **x summer** function evaluates the sum of all sequence and all magnitude-squared sequence values. In other words, it computes the sums

$$\sum_{n=-\infty}^{\infty} x[n] \qquad \text{and} \qquad \sum_{n=-\infty}^{\infty} |x[n]|^2$$

where $x[n]$ is a finite length input. The function prints the numerical values of these summations to the screen. It is executed by typing

$x[n]$

x summer $\overbrace{(\text{input})}$

x sview2

The **x sview2** function that stands for *superimposed* view is executed by typing

$$\text{x sview2} \qquad (\text{input1}) \qquad (\text{input2})$$

It displays two real discrete-time sequences superimposed on the same coordinate axes. This function is only intended for real sequences. Once the plot is displayed on the screen, three commands are available. Pressing the "*g*" key allows you to turn the grid of tic marks on or off to suit your preference. The "*esc*" key will return you to the display option menu for the case of complex signals. It will simply terminate the display if the signal is real. The "*q*" key will terminate the display in either case and return you to DOS. This function is intended for use on CGA-, EGA-, and VGA-equipped systems.

x truncate

The **x truncate** function is executed by typing

$$\text{x truncate} \qquad \overbrace{(\text{input})}^{x[n]} \qquad \overbrace{(\text{output})}^{y[n]} \qquad L$$

The function truncates the input sequence $x[n]$ at a specified ending sample L by implementing the relation

$$y[n] = \begin{cases} x[n], & n \leq L \\ 0, & \text{otherwise.} \end{cases}$$

L must be less than or equal to the length of the input.

x upsample

The **x upsample** function can be executed by typing

$$\text{x upsample} \qquad \overbrace{(\text{input})}^{x[n]} \qquad \overbrace{(\text{output})}^{y[n]} \qquad M$$

It performs a 1-to-M upsampling operation on the input sequence $x[n]$. The output sequence is defined by

$$y[n] = \begin{cases} x[n/M], & n/M \text{ an integer} \\ 0, & \text{otherwise.} \end{cases}$$

The length of the output sequence is approximately M times the length of the input signal.

x view

The **x view** function is executed by typing

$$\textbf{x view} \qquad \text{(input)}$$

It plots a discrete-time sequence on the screen. If the sequence is complex, a display option menu will appear that will ask you to specify display of the magnitude, log magnitude, phase, real part, imaginary part, magnitude and phase, or real and imaginary parts. Once the plot is displayed on the screen, three commands are available. Pressing the "*g*" key allows you to turn the grid of tic marks on or off to suit your preference. The "*esc*" key will return you to the display option menu for the case of complex signals. It will simply terminate the display if the signal is real. The "*q*" key will terminate the display in either case and return you to DOS. This function is intended for use on CGA-, EGA-, and VGA-equipped systems.

x view2

The **x view2** function is executed by typing

$$\textbf{x view2} \qquad \text{(input1)} \qquad \text{(input2)}$$

It is identical to the **x view** function except that it allows you to display two plots, one above the other.

x zeropad

The **x zeropad** function is executed by typing

$$\textbf{x zeropad} \qquad \overbrace{\text{(input)}}^{x[n]} \qquad \overbrace{\text{(output)}}^{y[n]} \qquad L$$

It appends zeros to the end of the input sequence $x[n]$. If $x[n]$ is non-zero in the range $N_1 \leq n \leq N_2$, then the output sequence is defined by

$$y[n] = \begin{cases} x[n], & N_1 \leq n \leq N_2 \\ 0, & N_2 < n \leq L \end{cases}$$

where L is the ending sample specified by the user.

Solution to Exercise 1.3.1

The macro **conjugate.bat** can be constructed by using the following commands:

```
x imagpart %1 junk
x gain junk junk 0 2
```

```
x subtract %1 junk %2
del junk
```

and is executed by typing

conjugate   input   output

Note that this will work on a complex sequence whether it is in Cartesian form or polar form.

Discrete-Time Signals and Systems

This chapter reviews some of the basic principles and definitions of digital signal processing. It uses several representations for signals, including functional expressions, graphs, and sequences of numbers. Pictorial representations of signals are particularly useful because they provide visual insight. Figure 2.1 graphically represents the sequence $x[n]$ as a function of the index n.

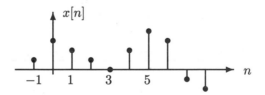

Figure 2.1. Pictorial representation of a sequence $x[n]$.

Equally important to a discussion of discrete-time signal processing is the concept of a discrete-time system. Section 2.2 discusses discrete-time systems and means for classifying them. Linear, time-invariant systems are highlighted because they have special properties that make them relatively easy to analyze, implement, and design.

2.1 DISCRETE-TIME SIGNALS

A number of elementary signals appear frequently in signal processing applications. Some of these are listed below and are pictured in Fig. 2.2.

- The *unit sample* or *unit impulse*

$$\delta[n] \triangleq \begin{cases} 1, & \text{if } n = 0 \\ 0, & \text{otherwise;} \end{cases}$$

- The *unit step*

$$u[n] \triangleq \begin{cases} 1, & \text{if } n \geq 0 \\ 0, & \text{if } n < 0; \end{cases}$$

- The *complex exponential*

$$e^{j\omega_0 n}$$

(ω_0 is called the *frequency*);
- The *sinusoid*

$$\sin[\omega_0 n + \theta_0]$$

(ω_0 is the frequency and θ_0 is the *phase offset*);
- The *one-sided exponential*

$$a^n u[n],$$

where $|a| < 1$ and $u[n]$ is the step function.

These elementary signals play an important role in describing more complex signals. For example, many useful signals can be expressed as weighted sums of damped exponentials or sinusoids. These and other elementary signals are examined in the exercises at the end of this section.

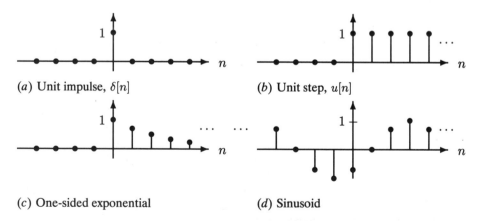

(a) Unit impulse, $\delta[n]$ (b) Unit step, $u[n]$

(c) One-sided exponential   (d) Sinusoid

Figure 2.2.   Some common signals.

The signal processing literature is filled with many terms that describe signals. These constitute a basic DSP vocabulary. These terms and acronyms are usually based on general characteristics of the signal. One such characteristic is whether the sequence is of *finite* or *infinite length*. A finite length sequence has a leftmost nonzero sample and a rightmost nonzero sample. The unit impulse is an example of a finite length sequence. An infinite length sequence extends infinitely far in one or both of the two directions. The unit step, exponential, and sinusoid are all examples of infinite length sequences.

Another useful signal characteristic is periodicity. A *periodic* sequence, $x[n]$, is infinite in duration and satisfies the relationship

$$x[n] = x[n + N] \quad -\infty < n < \infty$$

for some $N > 0$ called the *period*. The period N is always restricted to be a finite positive integer. Signals that are not periodic are called *aperiodic*. Aperiodic sequences may be of either finite or infinite duration and do not repeat indefinitely.

Signals can also be classified by whether their sample values are real or complex. Although most signals encountered in applications are real, it is often useful to work with complex signals in analyzing systems. Complex sequences can be represented in terms of their real and imaginary parts, $\Re e\{x[n]\}$ and $\Im m\{x[n]\}$, which are both real sequences, or in terms of their magnitudes and phases, $|x[n]|$ and $\angle x[n]$. For a complex sequence, $x[n]$, these are defined by

$$
\begin{aligned}
x[n] &\triangleq \Re e\{x[n]\} + j\, \Im m\{x[n]\} \\
&\triangleq |x[n]|e^{j\angle x[n]}
\end{aligned}
$$

The *complex conjugate* of $x[n]$ is the sequence $x^*[n]$.

$$
\begin{aligned}
x^*[n] &\triangleq \Re e\{x[n]\} - j\, \Im m\{x[n]\} \\
&\triangleq |x[n]|e^{-j\angle x[n]}
\end{aligned}
$$

Chapter 6 explores complex sequences in greater detail.

Signals may also be classified as *deterministic* or *random*. Deterministic sequences are sequences that can be expressed in a functional form. These include elementary signals such as sinusoids, steps, ramps, and the like, and arbitrary sequences composed of a weighted combination of elementary signals. Random signals, on the other hand, are probabilistic and "noise-like." The individual values in the sequence are less important than their distribution. Virtually all of the signals examined here are deterministic, but a few exercises at the end of this chapter and in the next focus on the generation of random signals and their properties.

Finally, signals may be classified according to their symmetry characteristics. Not all signals are symmetric, but all signals can be decomposed into even and odd symmetric sequences. Any arbitrary real sequence, $x[n]$, can be viewed as having two additive components: an *even part*, $x_e[n] = ev\{x[n]\}$, defined as

$$x_e[n] = \tfrac{1}{2}(x[n] + x[-n]);$$

and an *odd part*, $x_o[n] = od\{x[n]\}$, defined as

$$x_o[n] = \tfrac{1}{2}(x[n] - x[-n]).$$

The sum of the even and odd parts results in the original sequence. This concept is easily generalized to include complex signals of the form

$$x[n] = a[n] + jb[n]$$

where $a[n]$ and $b[n]$ are the real and imaginary parts, respectively. All complex sequences can be expressed as the sum of the four terms shown below

$$x[n] = \underbrace{a_e[n]}_{real-even} + \underbrace{a_o[n]}_{real-odd} +j \underbrace{b_e[n]}_{imaginary-even} +j \underbrace{b_o[n]}_{imaginary-odd} \qquad (2.1)$$

where the subscripts e and o denote the even and odd parts, respectively. Examining sequences in terms of their symmetry can sometimes greatly simplify their analysis and enable difficult problems to be solved in a simple way.

Summations involving sequences often arise in analyzing discrete-time systems. These summations can be expressed explicitly using the summation operator, $\sum$. However, in many cases these summations can be reduced to compact, closed-form expressions. Three of these can be derived from the geometric series:

$$\sum_{n=0}^{N-1} a^n = \frac{1 - a^N}{1 - a}, \qquad \forall \; a, \qquad (2.2)$$

where the symbol $\forall$ denotes "for all";

$$\sum_{n=0}^{\infty} a^n = \frac{1}{1 - a}, \quad \text{if } |a| < 1; \qquad (2.3)$$

and

$$\sum_{n=0}^{\infty} na^n = \frac{a}{(1 - a)^2}, \quad \text{if } |a| < 1. \qquad (2.4)$$

Equation (2.2) is valid for all complex values of the parameter a. Equation (2.3) is obtained by taking the limit of (2.2) as N goes to infinity. This is subject to the condition that $|a| < 1$. Equation (2.4) is related to equation (2.3) by differentiation. The exercises that follow explore some properties of sequences in more depth.

EXERCISE 2.1.1. **Shifting and Reversing Sequences**

Discrete-time signals are often expressed in terms of time-shifted and time-reversed combinations of other signals.

(a) The signal, $aa[n - N]$, is said to be *time shifted* by N where N is an integer. This means that the signal, $aa[n]$, is translated by N samples to the right if $N > 0$ and/or by $|N|$ samples to the left if $N < 0$. To illustrate this relatively simple concept, consider the following exercise. Begin by displaying the 5-point sequence, $aa[n]$, that may be found in the file **aa** using the **x view** function.

(i)   Sketch $aa[n]$.

(ii) Sketch $aa[n - 2]$.

(iii) Sketch $aa[n + 4]$.

Now check your sketches by using the computer. Use the **x lshift** function to perform the time shifting. Note that specifying positive integers shifts to the right; specifying negative integers shifts to the left.

(b) The signal, $aa[-n]$, is said to be *time reversed*. This means that the signal is flipped about the point $n = 0$.

(i)   Sketch $aa[-n]$.

(ii) If $v[n] = aa[n - 4]$, sketch $v[-n]$.

(iii) If $r[n] = aa[n + 2]$, sketch $r[-n]$.

Again check your sketches using the computer. Use the function **x reverse** to perform the time reversal.

EXERCISE 2.1.2.   Finite Length Sequences

The sequence

$$h[n] = u[n] - u[n - 10]$$

is a finite length sequence consisting of ones for n in the range $0 \leq n < 10$ and zeros everywhere else. It begins at $n = 0$ and ends at $n = 9$. Thus its length is 10 samples. On the other hand, the sequence

$$h[n] = a^n u[n]$$

is an example of an infinite length sequence, because it contains an infinite number of nonzero samples.

Several finite length sequences are listed below. Find the sequence lengths for each of these sequences and determine their starting and ending points. This exercise should be performed initially without the aid of the computer.

(a) $x[n] = u[n - 2] - u[n - 12]$

(b) $v[n] = u[n + 16] - u[n - 7]$

(c) $y[n] = x[n] \cdot v[n]$

(d) $r[n] = x[n] + v[n]$

(e) $s[n] = x[n+2] \cdot v[n-2]$

(f) $t[n] = y[n-1] + y[n+1]$

You can check your answers on the computer by using the **x siggen**, **x add**, **x subtract**, **x lshift**, and **x multiply** functions. For parts (a) and (b) the *block* option in **x siggen** can be used to approximate a step function.

Comment. It is important to note that a true step is infinite in duration, but a long block sequence can be used to approximate the step. A true step minus a shifted step results in a block or pulse. If steps are represented as finite length blocks, the difference between the block and shifted block produces erroneous sample values at the end of the sequence. To avoid this difficulty in this exercise, use block sequences of length 100 for the step functions. Before displaying your final result use **x truncate** with $L = 30$ to remove the erroneous samples at the end of the sequence.

EXERCISE 2.1.3. **Elementary Signals**

This problem explores a number of discrete-time signals that commonly arise in digital signal processing and provides a vehicle for becoming acquainted with the signal generator that will be used throughout this text. The signal generator can be executed by typing **x siggen**. It can generate a variety of different signals.

A small set of elementary signals is needed in this exercise. To begin, create the following signals of length 16 ($0 \le n \le 15$).

- $b[n]$, a 16-point block sequence with unit amplitude.
- $r[n]$, the first 16 points of the ramp function, defined as $nu[n]$.
- $t[n]$, 16 points of a periodic triangular wave with period 8, a maximum value of one, and starting point $n = 0$.
- $e[n]$, the first 16 points of the one-sided exponential, $(5/6)^n u[n]$.

Display and sketch them using **x view**. Signals are often formulated or expressed in terms of elementary signals. Using the elementary signals just created, sketch the following new signals:

(a) $v[n] = r[n-6]\, u[n]$;

(b) $a[n] = \begin{cases} b[n] - b[n-6], & n < 10 \\ 0, & \text{otherwise}; \end{cases}$

(c) $y[n] = e[n+10]\, b[n]$;

(d) $z[n] = t[n]\, (u[n] - u[n-10])$;

(e) $e_e[n] = ev\{e[n]\}\, (u[n+5] - u[n-5])$;

(f) $e_o[n] = od\{e[n]\}\, (u[n+5] - u[n-5])$.

This should be done initially without the aid of the computer, but you may use the functions **x lshift**, **x subtract**, **x truncate**, **x reverse**, and **x view** to check your results.

EXERCISE 2.1.4. **Time Axis Alteration**

The time axis of a signal, $x[n]$, can be altered to produce a new signal, $y[n]$. The method of alteration can be as simple as a time shift, or it can be a more complicated operation. A general time axis alteration can be expressed as

$$y[n] = x[f[n]]$$

where $f[n]$ is the time axis alteration function. The value of $f[n]$ must be an integer because the time index for digital sequences is only defined for integer values. As a simple example of time axis alteration, consider the problem of determining

$$y[n] = x[n - 6]$$

when

$$x[n] = u[n] - u[n - 5].$$

The sequence $x[n]$ is simply shifted to the right by six samples to produce the sequence $y[n]$. While this example poses no difficulty, determining the result of other alterations may not be so easy. For example, consider sketching

$$y[n] = x[3 - n]$$

where $f[n] = 3 - n$ and $x[n]$ is an arbitrary signal. A simple and systematic procedure for determining $y[n] = x[f[n]]$ can be summarized as follows:

1. Sketch $x[n]$.

2. Sketch another time axis underneath it for $y[n]$.

3. Write the expression for $f[n]$.

4. To find $y[0]$, evaluate $f[0]$ (which will always be an integer). Next evaluate $x[f[0]]$ and enter this value for $y[0]$ on the time axis for $y[n]$ at the point $n = 0$.

5. To find $y[1]$, evaluate $f[1]$ and $x[f[1]]$. Set $y[1] = x[f[1]]$ and enter it on the $y[n]$ time axis at the point $n = 1$.

6. Continue this procedure until all values, $-\infty < n < \infty$, of $y[n]$ have been found.

To test your understanding, consider the sequence $x[n]$ where

$$x[n] = \left(\tfrac{3}{4}\right)^n (u[n] - u[n - 5]).$$

(a) Without the aid of the computer, sketch $y[n]$ for each of the cases given below using the procedure just described.

   (i) $y[n] = x[-n]$.

   (ii) $y[n] = x[-7 - n]$.

   (iii) $y[n] = x[-n + 15]$.

(b) Develop a rule based on using the **x lshift** and **x reverse** operations that can be used to obtain $y[n]$ for each of the cases in part (a). Use **x siggen** to generate the damped exponential sequence segment, $x[n]$. Write a macro (as described in Section 1.3) using the **x reverse** and **x lshift** functions that will produce $y[n]$ for each of the cases in part (a). Indicate the specific shift parameter values that must be used in order for your macro to work properly. Check to verify that the answers obtained manually agree with those produced by your macro.

(c) Manually sketch the sequence

$$y[n] = x[2n].$$

Check your answer by using the function **x dnsample**. By selecting $M = 2$ in this function, you will be able to generate $y[n]$.

(d) Sketch the sequence

$$y[n] = x[5 - 2n].$$

EXERCISE 2.1.5. **Working with Summations**

Several useful closed-form expressions for summations exist, three of which are given in equations (2.2–2.4). Summations, like integrals, can be changed to alternate, but equivalent, forms through change-of-variables operations. This exercise should be performed analytically (without the aid of the computer).

(a) Consider the expressions shown below:

$$a_1 = \sum_{n=0}^{\infty} \left(\tfrac{1}{2}\right)^n$$

$$a_2 = \sum_{n=10}^{\infty} \left(\tfrac{1}{2}\right)^{n-10}$$

$$a_3 = 2^5 \sum_{n=5}^{\infty} \left(\tfrac{1}{2}\right)^n.$$

(i)   Determine the numerical values of a_1, a_2, and a_3.

(ii)   Show by using a change of variables in the summation index or by another approach that these summation expressions are related.

(b) Now consider these finite sums:

$$b_1 = \sum_{n=0}^{10} \left(\tfrac{1}{2}\right)^n$$

$$b_2 = \alpha_1 \sum_{n=-10}^{0} 2^n$$

$$b_3 = \alpha_2 \sum_{n=6}^{16} \left(\tfrac{1}{2}\right)^n$$

$$b_4 = \alpha_3 \sum_{n=0}^{10} \left(\tfrac{1}{2}\right)^{n+2}.$$

Determine the values for α_1, α_2, and α_3 so that

$$b_1 = b_2 = b_3 = b_4.$$

EXERCISE 2.1.6. The Geometric Series

Many problems become straightforward if the mathematical equations used in the problem descriptions are simplified. The signals shown below are expressed in terms of summations. Determine a simplified expression for each sequence so that it can be computed easily. Display and sketch each of these signals for the range $0 \le n \le 20$. Give the numerical values for the last three signal values (i.e., for $n = 18, \ 19, \ 20$) in each case. (*Hint:* Use the geometric series expressions to transform the signals into a convenient form.) Then use the computer to generate the final sequences using **x siggen, x add, x gain**, and **x multiply**. Examine the requested numerical values by either using a text editor or by typing the signal files to the screen.

(a)

$$x[n] = \left[\sum_{\ell=0}^{n} \left(\tfrac{7}{8}\right)^{\ell} - \sum_{\ell=0}^{n} \left(\tfrac{3}{4}\right)^{\ell} \right] u[n].$$

(b)

$$x[n] = \left[\sum_{\ell=0}^{\infty} (\ell+1)\left(\tfrac{1}{2}\right)^{\ell} + \sum_{\ell=0}^{n} \left(\tfrac{8}{9}\right)^{\ell} \right] u[n].$$

(c)

$$x[n] = \left[\sum_{\ell=-10}^{n-10} \left(\tfrac{7}{8}\right)^{\ell} \right] u[n].$$

(d)

$$x[n] = \left[\sum_{\ell=0}^{n} \ell \, (0.9)^{\ell} \right] u[n].$$

Part (d) may require some careful thought.

EXERCISE 2.1.7. **Characteristics of Signals**

In the introductory discussion, several common signal characteristics were defined. These included periodicity, signal duration, and the disposition of the sequence values with respect to being purely real, purely imaginary, or complex.

(a) Consider the signal

$$x[n] = \cos(\omega_0 n).$$

  (i)   Assume $\omega_0 = \pi/8$. Is this signal periodic? If so, determine its period and sketch one period of $x[n]$.

  (ii) Now let $\omega_0 = 1$. Is the discrete-time signal, $x[n]$, periodic? Now consider the continuous-time signal,

$$x(t) = \cos t.$$

What is the period of this signal? Explain why the discrete-time and continuous-time cosine signals behave differently.

(b) The signal

$$x[n] = e^{j\omega_0 n}u[n]$$

is infinite in duration. It is also complex valued. Generate a 40-point segment of the sequence, $x[n]$, where $\omega_0 = \pi/10$. This can be done by using the *exponential $Ke^{\alpha n}$* option in **x siggen** with $K = 1, alpha = 0 \ \ \pi/10$, and starting point at zero. Display and sketch the real and imaginary parts and the magnitude and phase of $x[n]$ using **x view**. Note that the real and imaginary parts are cosine and sine functions. This illustrates the relationship which is attributed to Euler:

$$e^{j\omega_0 n} = \cos \omega_0 n \ + \ j \sin \omega_0 n.$$

(c) Using the 40-point signal $x[n]$ defined in part (b), manually sketch each of the signals shown below:

$$y_0[n] = |10 \, x[n]|$$
$$y_1[n] = \angle x^*[n]$$
$$y_2[n] = \Re e\{jx[n]\}$$
$$y_3[n] = \Im m\{(jx[n])^*\}.$$

Now verify your results using the computer. The **x view** function will allow you to display the real and imaginary parts as well as the magnitude and phase. The **x gain** function (with *gain* $= 0 \ \ 1$) can be used to generate $jx[n]$; the **x conjugate** function will generate $x^*[n]$.

2.2 DISCRETE-TIME SYSTEMS

A discrete-time system can be viewed as a set of manipulations performed on one or more sequences. This text will primarily consider single input single output systems, since these represent the most common systems in traditional digital signal processing applications. Figure 2.3 depicts a typical system with input $x[n]$ and output $y[n]$. The input/output relationship can be written as $y[n] = T\{x[n]\}$ where $T\{\cdot\}$ denotes the system operation. Systems are commonly discussed and classified in terms of specific properties, some of which are now defined:

- *Additivity.* A system is *additive* if

$$T\{x_1[n] + x_2[n]\} = T\{x_1[n]\} + T\{x_2[n]\}$$

  for all real and complex inputs $x_1[n]$ and $x_2[n]$.

- *Homogeneity.* A system is *homogeneous* if

$$T\{\alpha x[n]\} = \alpha T\{x[n]\}$$

  for all complex values of α, and for all $x[n]$.

- *Linearity.* A system is *linear* if it is both additive and homogeneous.

- *Time Invariance or Shift Invariance.* A system is *time invariant* if a time shift in the input results in time shifting the output by the same amount. In other words, given that

$$y[n] = T\{x[n]\}$$

  a time-invariant system must satisfy the condition

$$y[n - n_0] = T\{x[n - n_0]\}$$

  for all integers, n_0, and all inputs, $x[n]$.

- *Stability.* A system is *unstable* if there exists a bounded input ($|x[n]| < \infty$ for all n) that causes the output signal $y[n]$ to be unbounded, i.e., $|y[n]| = \infty$. Otherwise the system is stable.[1]

- *Causality.* A system is *causal* if the output at time n is not dependent on future input samples. In other words, in a causal system output samples are only obtained from past or present samples of the input. Future input samples cannot be required to compute the output. For example, the system represented by the difference equation

$$y[n] = x[n] + x[n + 1]$$

  is not causal since $x[n + 1]$ is a future sample of the input. To see this clearly, assume that the present time is $n = 1$. Then $y[1] = x[1] + x[2]$. We see that $y[1]$

---

[1]This is called bounded input bounded output (BIBO) stability. There are other definitions of stability, but they will not be treated in this text.

requires that the present value $x[1]$ and the future value $x[2]$ be known. The system

$$y[n] = x[n] + x[n-1],$$

on the other hand, is causal because $x[n]$ and $x[n-1]$ represent present and past input samples, respectively.

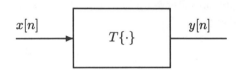

Figure 2.3. Block diagram representation of a system with input $x[n]$ and output $y[n]$.

Systems that are both linear and time invariant (LTI) are of special significance because they are well understood and can be conveniently analyzed in both the time domain and the frequency domain. The following exercises are designed to help you classify several unknown systems in terms of their properties. These systems, **system1**, **system2**, and **system3**, are provided as part of the software. Execution of these systems is best illustrated by an example. To execute **system1**, for example, you should simply type

system1   inputfile   outputfile

EXERCISE 2.2.1. **Testing Linearity**

Here you will work with an unknown system $T\{\cdot\}$. Use **x siggen** to create a ramp sequence $x_1[n]$ with a duration of five points in the range $0 \le n < 5$. Then create a 5-point sequence of ones, $x_2[n]$, also in the range $0 \le n < 5$ using the *block* option in **x siggen**. For the first part of this exercise, use **system1** as the unknown system.

- Generate $y_1[n] = T\{x_1[n]\}$.
- Generate $y_2[n] = T\{x_2[n]\}$.
- Generate $y_3[n] = T\{ax_1[n] + bx_2[n]\}$ by using the **x gain** and **x add** functions. Select arbitrary but convenient values for a and b.

(a) Examine the sequences $ay_1[n]$ and $by_2[n]$ for the same values of a and b. Does **system1** appear to be

(i) homogeneous?

(ii) additive?

(iii) linear?

Explain your reasoning.

(b) Now repeat this exercise using **system2** as the system $T\{\cdot\}$.

Note that if a counterexample is found that violates the linearity tests, one can be certain that the system is nonlinear. Proving that a system is linear based on this kind of probing is not practical since the definition requires that the linearity conditions be satisfied for all possible inputs.

EXERCISE 2.2.2. Testing Time Invariance

This exercise attempts to determine if a system $T\{\cdot\}$ is time invariant by observing the system's response to shifted inputs. Use a 5-point ramp sequence, $x[n]$, as the input and the unknown system, **system3**, as the system to be tested. It can be implemented by typing

$$\textbf{system3} \qquad \textbf{inputfile} \qquad \textbf{outputfile}.$$

Compute $y_1[n] = T\{x[n]\}$. Then use the **x lshift** function to generate the sequence $x[n-1]$. Finally, compute $y_2[n] = T\{x[n-1]\}$, using **system3**.

(a) What is your conclusion about the time invariance of this system?

(b) Repeat this problem using **system1** as the unknown system. Can you prove that **system1** is either time invariant or time varying based on your experimentation?

EXERCISE 2.2.3. Linear Time-Invariant (LTI) Systems

The trial-and-error method that was used in the previous exercises can be effective in determining if a system is nonlinear or time varying if a counterexample can be identified. However, it is very difficult to prove that an unknown system is linear or time invariant following this approach. If the equation that describes the system is known, we can determine with certainty whether a system is LTI. Based on the definitions, determine if the following systems with real inputs $x[n]$ and outputs $y[n]$ are: (1) linear and (2) time invariant.

(a) $y[n] = \sqrt{x[n] \cdot x[n]}$.

(b) $y[n] = x[\sqrt{n^2}]$.

(c) $y[n] = \log_2 10^{x[n]}$.

(d) $y[n] = \cos(x[n]) + \cos n$.

(e) $y[n] = x[n] + x[n+1] + x[n-2]$.

EXERCISE 2.2.4. Stability

Unstable systems are characterized by the ability to produce output sample values of infinite amplitude for bounded inputs. In practice, values of infinity cannot be represented, but an unstable system will often produce an output that saturates. This, in turn, leads to erroneous outputs if processing is allowed to continue. The system

$$y[n] = \frac{1}{x[n]}$$

is reasonably well behaved except when the numerical value of $x[n]$ is close to zero. When $x[n] = 0$, the output is clearly infinite and therefore the system is unstable.

Consider the system described by the relationship

$$y[n] = e^{x[n]}.$$

(a) Is this system stable?

(b) What numerical difficulties might be encountered in using this system?

(c) What is the approximate limit on the amplitude of the input for this system to operate properly on your machine? Check this experimentally using the function **x nlinear** to generate the signal $e^{x[n]}$. One way to do this is to generate a short *block* sequence ($x[n] = 1$) using **x siggen**. Use **x gain** to increase the amplitude of $x[n]$ until $e^{x[n]}$ saturates, i.e., results in an overflow error. You may wish to begin by considering the range $600 < x[n] < 800$.

EXERCISE 2.2.5. **Stability of LTI Systems**

For an LTI system with the impulse response, $h[n]$, a simple stability criterion exists. If

$$\sum_{n=-\infty}^{\infty} |h[n]| < \infty$$

then the LTI system is stable. The impulse responses for several LTI systems are listed below. Determine if they are stable.

(a) $h[n] = 2^n u[n]$.

(b) $h[n] = 2^n u[-n]$.

(c) $h[n] = (10n)^{-1} u[n - 1]$.

(d) $h[n] = u[n] - u[-n]$.

(e) $h[n] = 6^{-|n|}$.

EXERCISE 2.2.6. **More on Stability**

In this exercise, you will determine if a system is stable by examining the equation that defines the system. The system input and output are $x[n]$ and $y[n]$, respectively. For each of the systems shown below, analytically determine if the system is stable or unstable. Note that some of these systems are not LTI. This exercise does not require the use of the computer.

(a) $y[n] = x[n - 1] \cdot x[n]$.

(b) $y[n] = x[n-1]/x[n]$.

(c) $y[n] = \sum\limits_{k=0}^{\infty} x[n-k].$

(d) $y[n] = \sum\limits_{k=1}^{\infty} x[n-k]/k.$

(e) $y[n] = |x[n]|^2.$

(f) $y[n] = \sin(x[n]).$

EXERCISE 2.2.7. Causality

Causality is important in many practical systems, but it is critical in *real-time* systems. These are systems in which the input samples arrive sequentially and for which one output sample must be produced immediately after each input sample is received. These systems must be causal because only present or past input samples are available to be used to obtain the current output sample. This exercise does not require the use of the computer.

(a) To appreciate the importance of causality, consider the LTI system

$$y[n] = x[n] + 2x[n-1] + 3x[n-2].$$

Assume that the input signal is zero for $n < 0$ and calculate the system output step by step by using the following steps:

(i)   The first input sample (at time $n = 0$) is 1. What is the output $y[0]$?
(ii) The second input sample ($n = 1$) is 2. What is the output $y[1]$?
(iii) The next input sample ($n = 2$) is 3. What is the output $y[2]$?

Clearly, there is no problem in determining the value of each output sample as each input sample is received. This is because this system is causal. By contrast, consider repeating this exercise for the system

$$y[n] = x[n] + 2x[n+1] + 3x[n+2] + 4x[n+3].$$

It should be clear that it is not possible to determine $y[0]$ at time $n = 0$, $y[1]$ at time $n = 1$, etc., because all of the quantities on the right side of this equation are not available when they are needed.

(b) A way to circumvent the causality problem is to introduce a system delay. In particular, consider the same noncausal system of part (a) with a three sample system delay, i.e.,

$$y[n] = x[n-3] + 2x[n-2] + 3x[n-1] + 4x[n].$$

This is identical to the earlier system except that the output appears three samples delayed in time. Moreover, note that the new system is causal. If the input to the system is $x[n] = u[n] - u[n-4]$, determine the first four samples of the output analytically.

(c) The idea of using a system delay to make a noncausal system causal as discussed in part (b) has limitations. For example, if

$$y[n] = x[n]x[-n]$$

it is not possible to find a finite length delay to implement this system as a causal one. Find another noncausal system for which a causal implementation cannot be found by using a system delay.

2.3 CONVOLUTION

The *impulse response* of a system is the response of that system to a discrete-time impulse or unit sample. The impulse response has particular significance when a system is LTI because it completely characterizes the system. The impulse response provides all of the information that is necessary to determine the output of an LTI system in every situation. The output, $y[n]$, of an LTI system is given by the convolution of the input, $x[n]$, with the system impulse response, $h[n]$. This operation can be written as the *convolution sum*,

$$y[n] = \sum_{m=-\infty}^{\infty} x[m]\, h[n-m]. \tag{2.5}$$

It is common to use the shorthand notation

$$y[n] = x[n] * h[n] \tag{2.6}$$

where the asterisk "$*$" denotes (linear) convolution.

Convolution appears extensively in signal processing. While it may be viewed as a description of the input-output relationship for an LTI system, it can also be viewed as an operation performed between two arbitrary signals. Using this viewpoint, convolution can be used as a tool for analyzing the properties of signals or the processes that produced those signals.

Although it is not difficult to write a program to evaluate the convolution sum explicitly, an alternate method called *graphical convolution* can provide valuable insight into the convolution process. Using this approach, it is often possible to visualize the result of a convolution by inspecting the input sequences. The method involves properly aligning the two sequences on a common time axis, multiplying them, and then adding sequence values. The following exercises introduce convolution.

EXERCISE 2.3.1. **The Convolution Sum**

The definition of convolution is given in equation (2.5).

(a) Write a program, in any language that is supported in your computing environment, that will read in two sequences, convolve them, and write out the result.

Use the signal format described in Section 1.2 in your program so that it will be compatible with the other functions that are provided. Try to make the program as short and efficient as possible. Only a few lines of code should be needed for the program.

(b) The function **x convolve**, which is provided, also performs linear convolution. Generate two ramp functions using **x siggen**, one of length 15 and the other of length 10. Convolve these two sequences and provide a sketch of the two inputs and the output. Now use your program from part (a) to perform the same convolution and verify that your program works correctly.

EXERCISE 2.3.2. **An Interpretation of Convolution**

This exercise presents another interpretation of the convolution operation. Carefully examine the convolution sum and observe that

$$\delta[n] * x[n] = x[n],$$

$$\delta[n-1] * x[n] = x[n-1],$$

and

$$\delta[n-2] * x[n] = x[n-2].$$

Generalizing this observation, it follows that

$$\delta[n-n_0] * x[n] = x[n-n_0]$$

where n_0 is an arbitrary integer. Now consider the simple convolution $y[n] = aa[n] * x[n]$ where

$$x[n] = \delta[n] + \delta[n-1] + \delta[n-2] + \delta[n-3].$$

(a) The sequence $aa[n]$ is provided for you in the file **aa**. Use the *block* option of the function **x siggen** to generate $x[n]$. Sketch and display both of these sequences.

(b) Now create the four individual signals, $\delta[n]$, $\delta[n-1]$, $\delta[n-2]$, and $\delta[n-3]$ using **x siggen**. This can be done using the *block* option with different starting locations for the length 1 sequences. Sketch these sequences.

(c) Using the **x convolve** function create the sequence $y[n] = aa[n] * x[n]$ and sketch the result.

(d) Again using the **x convolve** function, convolve $aa[n]$ with each of the four sequences that you generated in part (b). Sketch these individual sequences.

(e) Add the resulting sequences from part (d) together using the **x add** function and sketch the result. Use **x view2** to compare the result with the result of part (b). Observe that the sum is identical to the sequence obtained by convolution.

The important point here is that any input sequence can be represented as a superposition of delayed unit impulses. The output is then equal to the same superposition of delayed unit impulse responses.

EXERCISE 2.3.3. **Introduction to Graphical Convolution**

Graphical convolution is a straightforward and intuitive procedure to evaluate the convolution sum without the aid of a computer. The method is motivated by recognizing that the term $h[n - m]$ in the convolution summation (2.5) can be interpreted as a signal in the variable m with n interpreted as a shift parameter. Since $h[n - m] = h[-(m - n)]$, it is a time-reversed copy of $h[m]$ that is then shifted by n samples on the m-axis. Thus, given $h[n]$, $h[n-m]$ can be obtained by first reflecting the signal on the m-axis about the point $m = 0$ to form $h[-m]$ and shifting it $|n|$ places to the left if n is negative or n places to the right if n is positive. This is illustrated in Fig. 2.4. The convolution at sample n can now be found by displaying $x[m]$ and $h[n-m]$ on a common m-axis, multiplying the two aligned sequences on a point-for-point basis, and summing the products. This exercise will walk through this procedure one step at a time.

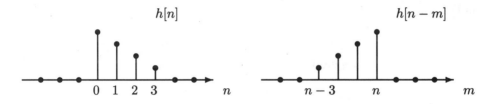

Figure 2.4.   The reflected impulse response of the convolution sum expressed as a sequence.

(a) First examine the sequences $x[n]$ and $h[n]$ that are provided in the files **bb** and **cc**, respectively, using **x view**. As you go through the procedure for the first time, sketch each of the shifted signals on graphs that are aligned vertically.

    (i) Reflect the sequence $h[n]$ about the origin using the **x reverse** function. Display and sketch your results. This is the signal $h[-m]$.

    (ii) To find $h[n_0 - m]$ at an arbitrary value $n = n_0$, shift the reflected sequence n_0 places to the right for $n_0 > 0$ or $|n_0|$ places to the left for $n_0 < 0$. For this example, begin by finding the value of $y[2]$, i.e., $n_0 = 2$. Use the **x lshift** function to slide the reflected signal two places to the right. Display and sketch your result.

    (iii) Now that the two sequences $h[m]$ and $x[2 - m]$ are aligned, multiply them pointwise using the **x multiply** function. Display and sketch your result.

    (iv) Add all the sample values in the product sequence together and assign that value to $y[n_0]$ (in this case $y[2]$). Do this step manually by printing the signal file to the screen. For this example, the correct value is -11. Therefore on a separate sheet of paper, record $y[2]$ graphically with a height of -11.

(v)   Repeat this procedure for all values of n, $-\infty < n < \infty$. In practice it is obviously not necessary to do this an infinite number of times when finite duration sequences are being convolved. Observe that shifting the reflected sequence too far to the left or right results in nonoverlapping sequences, in which case the result of the multiplication is an all-zero sequence. It is typical to start with the shift corresponding to the leftmost nonzero value of the output and increment n until the convolution is complete.

(b)   Given this step-by-step illustration of the procedure, determine and sketch the complete sequence $y[n]$. Observe that $y[n]$ is zero outside the range $0 \le n \le 7$. Moreover, observe that the total length of the output sequence is one less than the sum of the lengths of the two sequences that are convolved. Use the function **x convolve** to check your answer.

(c)   Determine and sketch $y[n]$ using graphical convolution by reflecting $x[n]$ instead of $h[n]$. Verify that your result is the same.

After this procedure becomes familiar, you should be able to reflect and align the sequences mentally to obtain a quick estimate of output values for simple convolutions.

EXERCISE 2.3.4. **Convolving Sequences**

In this exercise, discrete-time convolution is examined by example. Generate the following sequences:

$$x_1[n] = u[n] - u[n-4],$$
$$x_2[n] = u[n] - u[n-6],$$
$$x_3[n] = \left(\tfrac{7}{8}\right)^n (u[n] - u[n-5]),$$
$$x_4[n] = \sin\left(\tfrac{\pi n}{12} + \tfrac{\pi}{4}\right)(u[n] - u[n-6]).$$

Notice that $x_1[n]$ and $x_2[n]$ are 5-point and 7-point blocks, respectively, and that they can be created using **x siggen**. The signals $x_3[n]$ and $x_4[n]$ can be created using options 5 and 3, respectively. Test your intuition by performing the following convolutions in your head using the graphical convolution method.

(a) $y_1[n] = x_1[n] * x_2[n]$;

(b) $y_2[n] = x_1[n] * x_3[n-3]$;

(c) $y_3[n] = x_1[n] * x_4[n]$;

(d) $y_4[n] = x_3[n] * x_4[n]$.

In each case provide a rough and quick sketch. How long are the resulting sequences? Then check the accuracy of your result by performing the convolution operations using **x convolve** and **x lshift**.

EXERCISE 2.3.5. **More Convolution**

Consider the sequences shown below:

$$v_1[n] = \sum_{m=0}^{2} \delta[n - 6m]$$

$$v_2[n] = \sum_{k=1}^{3} \sum_{m=0}^{2} \delta[n - 6m - k]$$

$$v_3[n] = \left(\tfrac{2}{3}\right)^n (u[n] - u[n - 5]).$$

(a) Use **x siggen** to create these sequences. This can be done easily by using options 8, 1, and 5. Alternatively, in the case of $v_s[n]$ you may wish to convolve a block signal with an impulse train. Display and sketch each result.

(b) Using graphical convolution, sketch the following sequences as accurately as you can.

(i) $y_0[n] = v_1[n] * v_3[n]$;

(ii) $y_1[n] = v_1[n] * v_1[n] * v_1[n]$;

(iii) $y_2[n] = v_1[n] * v_2[n]$;

(iv) $y_3[n] = v_3[n] * v_2[n]$;

(v) $y_4[n] = v_3[n] * v_3[n - 1]$.

(*Note:* For this case your sketch will probably be rather rough. However, you should be able to capture the general shape of the output.)

Check your results using the computer.

EXERCISE 2.3.6. **Beginning and End Points**

The convolution of two finite length sequences always results in a sequence of finite length. Graphical convolution can be used to determine when the output sequence begins and ends.

This exercise will use the following sequences:

(i) $x[n] = u[n] - u[n - 50]$;

(ii) $v[n] = u[n - 31] - u[n - 42]$;

(iii) $w[n] = u[n + 51] - u[n + 17]$.

(a) Determine the first and last nonzero samples for each of the following:

(i) $y_1[n] = x[n] * v[n]$;

(ii) $y_2[n] = v[n] * w[n]$;

(iii) $y_3[n] = x[n] * w[n]$.

Use the computer to check your results. Derive a rule for determining the initial and final samples for the convolution of two finite length sequences.

(b) Determine the initial and final samples for the following sequences without using the computer:

(i) $y_4[n] = x[n] * z[n]$;

(ii) $y_5[n] = v[n] * z[n]$;

(iii) $y_6[n] = z[n] * w[n]$.

where $z[n] = u[n + 140820] - u[n - 213]$. Note that the length of $z[n]$ exceeds the signal length limit of the software. Hence you cannot use the computer to get the correct answer.

EXERCISE 2.3.7. Convolution of Infinite Length Sequences

Consider the sequences

$$x_1[n] = \left(\tfrac{1}{2}\right)^n u[n]$$

$$x_2[n] = 2^n u[-n]$$

$$x_3[n] = u[n] - u[n - 10].$$

Using graphical convolution, sketch the general shape of the following infinite length signals.

(a) $y_a[n] = x_1[n] * x_1[n]$.

(b) $y_b[n] = x_2[n] * x_2[n]$.

(c) $y_c[n] = x_2[n] * x_3[n]$.

(d) $y_d[n] = x_1[n] * x_2[n]$.

You can obtain an approximate check of your answer using **x siggen** to generate finite length approximations (of 20 samples or so) to these signals and then convolving them together using **x convolve**. To generate $x_2[n]$ first create 20 samples of $(1/2)^n u[n]$ using **x siggen** and then use **x reverse**.

EXERCISE 2.3.8. Using Graphical Convolution

In this exercise, you are to obtain the value of the convolution of two sequences at one specific sample location. Consider the sequences

(i) $x[n] = u[n] - u[n - 47]$;

(ii) $v[n] = u[n + 28] - u[n - 10]$;

(iii) $w[n] = \left(\tfrac{3}{4}\right)^n u[n]$;

(iv) $r[n] = \left(\tfrac{3}{4}\right)^{|n|}$;

(v) $z[n] = (n + 12)\left(u[n + 12] - u[n - 30]\right)$.

Using the graphical convolution procedure and without the aid of the computer, determine the values as indicated below.

(a) If $y_a[n] = x[n] * v[n]$, find $y_a[10]$ and $y_a[43]$.

(b) If $y_b[n] = x[n] * w[n]$, find $y_b[40]$ and $y_b[60]$.

(c) If $y_c[n] = v[n] * r[n]$, find $y_c[-29]$ and $y_c[-2]$.

(d) If $y_d[n] = w[n] * z[n]$, find $y_d[12]$.

EXERCISE 2.3.9. Algebraic Properties of Convolution

Several properties of convolution that are often exploited in discrete-time signal processing are examined here. These properties will be examined by using the sequences $x[n]$ and $h[n]$, where

$$x[n] = n \; (u[n] - u[n-32])$$
$$h[n] = 2n \; (u[n] - u[n-26]).$$

(a) Show analytically that $y[n] = x[n] * h[n] = h[n] * x[n]$ by changing the variables in the summation in equation (2.5). Then verify that $x[n] * h[n] = h[n] * x[n]$ by generating these sequences with **x siggen** and using the **x convolve** function. The convolution operation is said to be *commutative*. Sketch $y[n]$.

(b) Now split $x[n]$ into two sequences of the same length (32 points), called $x_1[n]$ and $x_2[n]$, such that $x[n] = x_1[n] + x_2[n]$. Provide a sketch of these two sequences. Note that there are many choices for $x_1[n]$ and $x_2[n]$ that satisfy this condition. A simple way to do this is to use **x siggen** to design two 32-point ramp functions (perhaps with different slopes) that add to $x[n]$. Compute and sketch $y_1[n] = x_1[n] * h[n]$, $y_2[n] = x_2[n] * h[n]$, and $y[n] = y_1[n] + y_2[n]$. Observe that $y[n]$ is exactly the same as $x[n] * h[n]$ computed in part (a).

(c) Repeat the exercise performed in part (b), but with the condition that $x_1[n]$ and $x_2[n]$ are of different lengths. One way to do this is to let $x_1[n]$ be a truncated version of $x[n]$ and $x_2[n]$ be $x[n] - x_1[n]$. Is the equal length condition a necessary requirement? Convolution is said to be *distributive* with respect to addition.

EXERCISE 2.3.10. System Impulse Response

Knowledge of the impulse response of a system is important in digital signal processing. Here you will find the impulse response of the system $T\{\cdot\}$ that is implemented in **system1**.

(a) Find and sketch the impulse response, $h[n]$, of the system $T\{\cdot\}$. Use the *block* option of **x siggen** to create the impulse.

(b) Compute and sketch $y_1[n] = T\{x_1[n]\}$ where $x_1[n] = 2\delta[n-3]$. Use the statement

$$\textbf{system1} \qquad \textbf{inputfile} \qquad \textbf{outputfile}$$

to implement **system1** and **x siggen** to generate $x_1[n]$. Compute and sketch $y_2[n] = x_1[n] * h[n]$ using **x convolve**.

(c) Do the sequences $y_1[n]$ and $y_2[n]$ suggest that $T\{\cdot\}$ is LTI? Explain.

2.4 DIFFERENCE EQUATIONS

There are many types of difference equations. The most common type in digital signal processing is the linear constant coefficient difference equation or LCCDE. These can be used to represent certain LTI systems. Assume that a system with input, $x[n]$, and output, $y[n]$, is defined by an LCCDE of the form

$$y[n] + a_1 y[n-1] + a_2 y[n-2] + \cdots + a_N y[n-N]$$
$$= b_0 x[n] + b_1 x[n-1] + \cdots + b_M x[n-M].$$

This equation can be rewritten as the recursion

$$y[n] = -a_1 y[n-1] - a_2 y[n-2] - \cdots - a_N y[n-N]$$
$$+ b_0 x[n] + b_1 x[n-1] + \cdots + b_M x[n-M]$$

where $a_1, a_2, \ldots, a_N$ and $b_0, b_1, \ldots, b_M$ are the constant coefficients. To specify the output uniquely we must also specify N initial sample values of $y[n]$. If the system is causal and its output samples are to be evaluated beginning at $n = 0$ (i.e., $y[n]$ is to be evaluated for $n = 0, 1, \ldots$), then the sample values

$$y[-1], y[-2], \ldots, y[-N]$$

must be prespecified. These are called the *initial conditions*. Once the initial conditions are given, the output samples can be computed sequentially. When these initial conditions are assumed to be zero, LCCDEs can define causal, linear, time invariant systems for inputs that are zero for $n < 0$. These difference equations are commonly used to implement discrete-time LTI systems. LCCDEs are discussed further in Chapter 5.

EXERCISE 2.4.1. **Evaluating a Difference Equation**

In this exercise you will evaluate the output of a causal LTI system specified by the simple difference equation

$$y[n] = \tfrac{1}{2}y[n-1] + x[n] + 2x[n-1]$$

with zero initial conditions. Assume that the input is

$$x[n] = \delta[n] + \delta[n-1].$$

A convenient way to evaluate the output samples without a computer is to arrange the computation in a chart.

n	$x[n]$	$x[n-1]$	$y[n-1]$	$y[n]$
$n = -1$	$x[-1] = 0$	$x[-2] = 0$	$y[-2] = 0$	$y[-1] = 0$
$n = 0$	$x[0] = 1$	$x[-1] = 0$	$y[-1] = 0$	$y[0] = 1$
$n = 1$	$x[1] = 1$	$x[0] = 1$	$y[0] = 1$	$y[1] = 3.5$
$\vdots$	$\vdots$	$\vdots$	$\vdots$	$\vdots$

Since this system is linear and causal, no nonzero output sample is produced until a nonzero input sample is received. The input in our example begins at $n = 0$. Therefore all samples in the output sequence will be zero until this time and the chart can begin at $n = 0$. On the line corresponding to $n = 0$, $x[0] = 1$ and $x[-1] = y[-1] = 0$. Substituting these values into the difference equation gives $y[0] = 1$. This procedure can be continued by going to the next line, entering the values for $y[0]$, $x[1]$, and $x[0]$, and then evaluating $y[1]$ using the difference equation and continuing similarly until all the desired samples of $y[n]$ have been computed.

(a) Following this procedure, fill in the lines for $n = 2$, 3, 4. The column in the chart labeled $y[n]$ is the system output.

(b) A computer is ideally suited to evaluate the output samples specified by a difference equation. Use **x siggen** to create the input, then use **x lccde** to compute the output. Record the first seven values of $y[n]$ by typing the output file to the screen.

EXERCISE 2.4.2. **Convolution Versus LCCDEs**

LCCDEs with zero initial conditions are often used to implement LTI systems. Consider the causal LCCDE

$$y[n] = \tfrac{2}{3}y[n-1] + x[n]$$

with input $x[n]$ and output $y[n]$.

(a) Determine the impulse response, $h[n]$, of this system by evaluating the difference equation using an impulse as the input and zero for the initial condition at $n = -1$. The function **x siggen** can be used to create the impulse and the function **x lccde** will implement the difference equation. The impulse response has the form

$$h[n] = \alpha^n u[n].$$

What is the value of α? Note that we can implement this system even though its impulse response is infinite in duration.

(b) In part (a) you found the impulse response and observed that it was infinitely long. If it can be assumed that all values of $h[n]$ less than 0.001 are negligible and can be ignored, what is the effective length of $h[n]$? You may find it more convenient to read the sequence amplitude values by typing the signal file to the screen or by using a text editor rather than by using **x view**. Generate this finite length version of $h[n]$ and the sequence

$$x[n] = u[n] - u[n-6]$$

using **x siggen**. Compute and sketch $x[n] * h[n]$.

(c) Now implement the convolution in part (b) explicitly using a difference equation. Use the **x lccde** function with zero initial conditions. Compare your results with the sequence obtained by convolution. By using the function **x subtract** display and sketch the difference between the two results. How many multiplies and adds are required on average to obtain each output sample using both of these methods?

EXERCISE 2.4.3. **The First Backward Difference**

The discrete-time first backward difference operation is similar to taking the derivative of a continuous-time signal in certain respects. It takes the difference between the current sample of $x[n]$ and the previous one, i.e.,

$$y[n] = x[n] - x[n-1]. \tag{2.7}$$

It corresponds to an LTI system with the impulse response

$$h[n] = \delta[n] - \delta[n-1]. \tag{2.8}$$

To explore this operation, this exercise will consider the effects of the first backward difference operation on several sequences. To begin use the *create file* option in **x siggen** to create a file containing $h[n]$.

(a) Sketch the continuous-time function,

$$\text{tri}_a(t) = t\,u(t) - 2(t-20)\,u(t-20) + (t-40)\,u(t-40).$$

Note that this is a continuous-time triangular function beginning at $t = 0$ and ending at $t = 40$, with a peak value at $t - 20$. In addition, sketch the continuous-time square pulse,

$$\text{sq}_a(t) = u(t) - u(t-20).$$

Now compute (by inspection) and sketch the first derivative of each of these continuous-time waveforms.

(b) Generate the 41-point triangular pulse

$$\text{tri}[n] = n\,u[n] - 2(n-20)\,u[n-20] + (n-40)\,u[n-40].$$

This can be done easily using the *triangular wave* option in **x siggen** by specifying the period to be 40 and the number of periods to be one. In addition, generate and sketch the square pulse

$$\text{sq}[n] = u[n] - u[n-20].$$

Use the *square wave* option in **x siggen** with pulse length 20, period 40, and number of periods 1. Now compute and sketch the first backward difference of each of these signals using either **x lccde** to implement the system using equation (2.7) or **x convolve**. Note the similarity to the continuous-time case. The function **x view2** may be used to display the signal and its first backward difference one above the other. In many cases the first backward difference is used as an approximation to a derivative operator.

EXERCISE 2.4.4. **The Running Sum**

The discrete-time running sum operation is in many ways analogous to continuous-time integration. The running sum of a sequence $x[n]$ is defined by

$$y[n] = \sum_{k=-\infty}^{n} x[k].$$

It is a linear, time-invariant operation that can be implemented as a difference equation. If the input is $x[n]$ and the running sum output is $y[n]$, then

$$y[n] = y[n-1] + x[n]$$

where zero initial conditions are assumed. This exercise will consider the performance of the running sum operation on several elementary sequences.

(a) Consider the continuous-time signals:

$$\mathrm{sq}_a(t) = u(t) - u(t-20),$$

$$\mathrm{imp}_a(t) = \sum_{k=0}^{6} \delta(t-10k),$$

and

$$\cos(\tfrac{\pi}{5}t).$$

Sketch each of these signals and also the signals that you would expect to obtain by integrating each of them.

(b) Generate the following discrete-time signals:

(i)   One period of a square wave,

$$\mathrm{sq}[n] = u[n] - u[n-20].$$

Do this by using the square wave option in **x siggen** with the pulse length specified to be 20 and the period 128.

(ii)   Six periods of the impulse train,

$$\mathrm{imp}[n] = \sum_{k=0}^{6} \delta[n-10k].$$

This can be done using the *impulse train* option in **x siggen** with period 10.

(iii) A sine wave of the form

$$\sin(\tfrac{\pi}{40}n).$$

Generate this sequence by using the *sine wave* option in **x siggen** with *alpha* $= \pi/40, phi = 0$ and length $= 128$.

Evaluate and sketch the running sum of each of these signals. Compare these results with the results found for the continuous-time case in part (a). You should observe that the running sum performs like a discrete-time integrator.

EXERCISE 2.4.5. **Nonlinear Difference Equations**

The class of nonlinear difference equations is extremely broad, but its use is often limited by the lack of a well-developed theory for nonlinear systems. This exercise explores a nonlinear difference equation that generates a Gaussian sequence of the form

$$f[n] = \exp(-\alpha(n\tau)^2) \tag{2.9}$$

where α is positive. For this exercise, assume $\alpha = 1$ and $\tau = 0.02$.

(a) Generate the Gaussian function described by equation (2.9) for $n = 0, 1, 2, \ldots,$ 99. Do this by using the ramp option in **x siggen** and the **x gain** function to generate $n\tau$, the **x nlinear** and **x gain** functions to form $-\alpha(n\tau)^2$, and **x nlinear** to perform the exponentiation. Sketch and display your results.

(b) Write a program for evaluating this same function using the nonlinear difference equation (or recursion)

$$f[n+1] = c\frac{f^2[n]}{f[n-1]}$$

where $c = \exp(-2\alpha\tau^2)$. This recursion is due to Teager (see Kaiser [1]). The two required initial conditions, $f[-1]$ and $f[0]$, can be obtained directly from the Gaussian function. Evaluate the function over the range $0 \le n \le 511$ and sketch the result. Your computer program may be written in any language that you choose.

(c) Kaiser [1] has shown that Teager's difference equation can be expressed as the following pair of first-order nonlinear difference equations

$$h[n+1] = c\, h[n]$$
$$f[n+1] = h[n+1]\, f[n].$$

To begin the recursion, a value for $h[0]$ is needed. Observe from the equation that $h[0] = f[0]/f[-1]$. The initial conditions for $f[n]$ can be obtained from equation (2.9). Write a program that implements this simple pair of equations and evaluate the Gaussian as in part (b). Compare the results in all three cases. What is the maximum error?

EXERCISE **2.4.6. Autocorrelation**

In some applications, such as radar signal processing, seismic analysis, and signal enhancement, an information-bearing signal is obscured by some form of noise. The goal is to estimate certain signal characteristics from the corrupted signal. This exercise will examine autocorrelation as a procedure for determining the period of a quasi-periodic sequence that is embedded in additive noise.

The autocorrelation of a real signal $v[n]$ is defined as

$$\mathcal{A}[n] = \sum_{m=-\infty}^{\infty} v[n+m] \, v[m].$$

It is related to the convolution of $v[n]$ with itself, since

$$\mathcal{A}[n] = v[n] * v[-n].$$

$\mathcal{A}[n]$ provides a measure of similarity for different alignments of the signal.

(a) Generate 128 points of the sequence $x[n]$, where

$$x[n] = \tfrac{1}{2} \sin\left(\tfrac{\pi}{12}n\right) + \sin\left(\tfrac{\pi}{24}n\right),$$

using the *sine wave* option in **x siggen** and the **x add** function. Next, generate 128 samples of a random noise sequence $r[n]$ using **x rgen**. Use **x gain** and **x add** to generate the sequence

$$y[n] = x[n] + 3r[n]$$

that will serve as the noisy signal. Examine $x[n]$ using the **x view** function and determine the period of this signal. Now examine $y[n]$ using **x view** and observe that it is difficult to determine the period by inspection.

(b) The autocorrelation function can be used to estimate the periodicity of the signal $x[n]$. In part (a) you determined the period of $x[n]$ by inspection. Compute the autocorrelation of $x[n]$ using **x reverse** and **x convolve** and sketch your results. Using the concept of graphical convolution, explain why peaks in the autocorrelation function appear at values of n that are multiples of the period.

(c) The autocorrelation function can be used to estimate the periodicity of the signal $x[n]$, given the noisy signal $y[n]$. Specifically, the distance between the peak at the origin and either of the adjacent peaks in the autocorrelation function gives an estimate of the period of $x[n]$. Display and sketch the autocorrelation function of $y[n]$ and estimate the period of the underlying signal $x[n]$.

EXERCISE **2.4.7. Cross Correlation**

Cross correlation can be viewed as a generalization of autocorrelation. The cross correlation of two real sequences $v[n]$ and $w[n]$ is defined as

$$\Phi[n] = \sum_{m=-\infty}^{\infty} v[n+m] \, w[m].$$

If $v[n]$ and $w[n]$ are equal, the cross correlation and autocorrelation are identical. Cross correlation also has a convolutional interpretation,

$$\Phi[n] = w[n] * v[-n].$$

It can be used to determine when two sequences are best aligned in time. For example, if $x[n]$ and $v[n]$ are two different but similar sequences, one may wish to determine which value of ℓ results in the best approximation

$$x[n] \approx v[n + \ell].$$

This can be done by computing the cross correlation and looking for the time index, n, of the maximum peak of $\Phi[n]$. To examine this point further, create the sequences $x[n]$ and $v[n]$ where $x[n]$ is a random finite length sequence that is nonzero in the range $-32 \le n \le 32$. Let

$$v[n] = x[n] * h[n]$$

where $h[n]$ is a 15-point lowpass filter with cutoff frequency of $2\pi/3$ and starting point zero. Use **x rgen** to create $x[n]$ and **x lshift** to shift the starting point to -32. The *Hamming window* option in **x fdesign** can be used to create $h[n]$.

(a) Use **x view2** to display and sketch $x[n]$ and $v[n]$. These signals are not the same but are similar in terms of their distribution of samples. Such sequences have a degree of correlation as defined by their cross correlation function. Now compute and sketch $\Phi[n]$. Identify the index n corresponding to the peak in $\Phi[n]$. To simplify the task of finding the peak location, use the **x extract** function to extract the 20-point block in $\Phi[n]$ beginning at -10. Write this block to a sequence file that also begins at -10 and display it. By examining the extracted sequence file, you will be able to determine the peak location by inspecting the display. This location provides an indication of how much displacement is necessary for $v[n]$ to be most nearly aligned with $x[n]$. Based on your examination of the peak location, by how many samples should $v[n]$ be shifted to produce the best alignment between the sequences? Check your answer by shifting $v[n]$ by that amount and visually comparing it to $x[n]$. Approximately how many multiplies and adds are required to compute the cross correlation in this example?

(b) In this part, a method for increasing the efficiency of computing the cross correlation function is examined. Quantize each of the sequences into binary sequences in the following way:

$$\hat{x}[n] = \begin{cases} 1 & \text{if} \quad x[n] \ge 0 \\ -1 & \text{if} \quad x[n] < 0 \end{cases}$$

$$\hat{v}[n] = \begin{cases} 1 & \text{if} & v[n] \geq 0 \\ -1 & \text{if} & v[n] < 0. \end{cases}$$

Use the **x quantize** function to do this with *min* $= -1$ and *max* $= 1$ and the number of levels equal to two. Compute the cross correlation function of $\hat{x}[n]$ and $\hat{v}[n]$. Use **x view2** to display this cross correlation function and the one in part (a) simultaneously. What is the required shift in $\hat{v}[n]$ to produce the best alignment using this approach? How many multiplies and adds are required? How do the methods in parts (a) and (b) compare? (*Note:* Multiplication by $+1$ or -1 should not be counted as a multiply.) You will not observe any speedup in the software that is provided because those programs do not check to avoid multiplications by ± 1.

(c) Two random signals generated with different seed values should not bear any similarity to each other and are said to be uncorrelated. Consequently, they should not produce a cross correlation function with a dominant peak indicating a preferred alignment. Use **x rgen** and **x lshift** as before to create another random sequence, $y[n]$, (with a different seed) in the same range. Display and sketch the cross correlation of $x[n]$ and $y[n]$. Observe that there is no dominant peak that clearly marks a "best alignment" between the two sequences.

EXERCISE 2.4.8. Random Sequences

It is often important to generate a random sequence of numbers that has a Gaussian distribution instead of a uniform distribution. There are many methods available for doing this. A simple one that transforms a uniform random sequence into a Gaussian random sequence is investigated here.

The central limit theorem states that the sum of N identically distributed, independent random variables approaches a Gaussian distribution in the limit as N goes to infinity.

(a) Generate a uniform random sequence $x_1[n]$ of length 128 using **x rgen**. Use the **x gain** function to multiply the sequence by 100 to produce a random sequence, $y[n]$, with amplitude values in the range -50 to 50. To examine the distribution of $y[n]$, compute and display its histogram using **x histogram** and **x view**. The histogram is a plot of sample amplitude on the x-axis versus the number of occurrences of that amplitude on the y-axis. If the histogram is roughly flat, the sequence is said to be uniformly distributed.

(b) Now generate eight additional random sequences $x_2[n]$, $x_3[n]$, $\ldots$, $x_9[n]$. Add the sequences $x_1[n], \ldots, x_9[n]$ together using **x add** to form $y_2[n]$. Multiply $y_2[n]$ by 11 so that its amplitude range will be from -50 to $+50$. Display and sketch the histogram of the new sequence. Does this distribution more closely resemble that of a Gaussian random sequence?

EXERCISE 2.4.9. Gaussian Random Sequences

This problem considers a different method for creating a Gaussian sequence [2] with a specific standard deviation σ. With this method, a uniform random sequence

$0 < x_1[n] \le 1$ is first transformed into a Rayleigh-distributed random sequence $r[n]$ by the relation

$$r[n] = \sqrt{2\sigma^2 \ln\left(\frac{1}{x_1[n]}\right)}.$$

Two uncorrelated (and therefore independent) Gaussian random variables, $g_1[n]$ and $g_2[n]$ can then be formed using the equations

$$\begin{aligned} g_1[n] &= r[n]\cos(2\pi x_1[n+1]), \\ g_2[n] &= r[n]\sin(2\pi x_1[n+1]). \end{aligned}$$

The standard deviation is the value of σ used to generate $r[n]$.

(a) Write a macro that will accept a uniform random sequence as its input and produce $g_1[n]$ and $g_2[n]$ as its outputs. The functions **x divide, x log, x multiply, x lshift,** and **x nlinear** will be helpful in computing $r[n]$. Use your macro to generate $g_1[n]$ and $g_2[n]$ with $\sigma = 1$ and length 1024. List all necessary functions and intermediate signal files in your macro. Note that the random signal $x_1[n]$ should have an amplitude range between 0 and 1. Therefore you should add a *block* sequence with amplitude .5 to the random sequence generated by **x rgen** in constructing $x_1[n]$.

(b) Determine the maximum absolute value A for $g_1[n]$ and $g_2[n]$. Multiply each by $1000/A$ so that the amplitude range of each signal will be between -500 and $+500$. Now compute the histogram of $g_1[n]$ and $g_2[n]$ and sketch the results. Does this distribution appear to be Gaussian?

(c) Cross correlation was introduced in Exercise 2.4.7. Compute the cross correlation function for $g_1[n]$ and $g_2[n]$. Do these sequences appear to be uncorrelated? Justify your conclusion.

2.5 REFERENCES

[1] J. F. Kaiser, "On the Fast Generation of Equally Spaced Values of the Gaussian Function $A \exp(-at * t)$," *IEEE Transactions on Acoustics, Speech, and Signal Processing*, Vol. ASSP-35, No. 10, pp. 1480–1481, Oct. 1987.

[2] L. Rabiner and B. Gold, *Theory and Application of Digital Signal Processing*, Prentice-Hall, Englewood Cliffs, NJ, 1975.

The Frequency Domain 3

3.1 THE DISCRETE-TIME FOURIER TRANSFORM

The concept of the frequency domain is as important for understanding discrete-time signals and systems as it is for understanding the continuous-time case. A careful examination of the spectrum of a signal provides important clues about how that signal should be analyzed. The frequency response of a system tells us how the system will respond to unknown, and possibly difficult to characterize, inputs. This chapter focuses on the discrete-time Fourier transform (DTFT) and the frequency response of discrete-time systems. These are the most important of the frequency-domain analysis tools. It also discusses a number of properties of the DTFT. The chapter concludes with a brief discussion of filtering. A more complete discussion of the latter topic can be found in the companion text *Digital Filters: A Computer Laboratory Text*.

The discrete-time Fourier transform, which serves as the Fourier representation of a discrete-time signal, is defined by the summation

$$X(e^{j\omega}) = \sum_{n=-\infty}^{\infty} x[n]e^{-j\omega n}. \tag{3.1}$$

It is a periodic function in the real variable ω with a period of 2π. The DTFT is equivalent to the representation of a signal as a sequence of samples, since the samples can be computed from the DTFT and vice versa. The integral

$$x[n] = \frac{1}{2\pi} \int_{-\pi}^{\pi} X(e^{j\omega})e^{j\omega n} \, d\omega \tag{3.2}$$

is used to evaluate the inverse DTFT. Note that the range of integration is limited to one period of $X(e^{j\omega})$. These two formulas resemble those that define the Fourier series representation of a periodic function. In fact, the sample values $x[n]$ can be identified with the Fourier series coefficients of $X(e^{j\omega})$. Table 3.1 contains an abbreviated list of discrete-time Fourier transform pairs.

Table 3.1.   A Short List of DTFT Pairs[a]

$x[n]$	$\Longleftrightarrow$	$X(e^{j\omega})$
1	$\Longleftrightarrow$	$2\pi\delta((\omega))_{2\pi}$
$e^{j\omega_0 n}$	$\Longleftrightarrow$	$2\pi\delta((\omega - \omega_0))_{2\pi}$
$\cos\omega_0 n$	$\Longleftrightarrow$	$\pi\delta((\omega - \omega_0))_{2\pi} + \pi\delta((\omega + \omega_0))_{2\pi}$
$\sin\omega_0 n$	$\Longleftrightarrow$	$\dfrac{\pi}{j}\delta((\omega - \omega_0))_{2\pi} - \dfrac{\pi}{j}\delta((\omega + \omega_0))_{2\pi}$
$\displaystyle\sum_{k=-\infty}^{\infty}\delta[n - kN]$	$\Longleftrightarrow$	$\dfrac{2\pi}{N}\displaystyle\sum_{k=-\infty}^{\infty}\delta\left(\left(\omega - \dfrac{2\pi k}{N}\right)\right)_{2\pi}$
$a^n u[n],\ \|a\| < 1$	$\Longleftrightarrow$	$\dfrac{1}{1 - ae^{-j\omega}}$
$x[n] = \begin{cases} 1, & \|n\| \le N \\ 0, & \|n\| > N \end{cases}$	$\Longleftrightarrow$	$\dfrac{\sin[\omega(N + \frac{1}{2})]}{\sin[\frac{\omega}{2}]}$
$\dfrac{\sin\alpha n}{\pi n}\qquad 0 < \alpha < \pi$	$\Longleftrightarrow$	$X(\omega) = \begin{cases} 1, & 0 < \|\omega\| < \alpha \\ 0, & \alpha < \|\omega\| \le \pi \end{cases}$
$\delta[n]$	$\Longleftrightarrow$	1
$u[n]$	$\Longleftrightarrow$	$\dfrac{1}{1 - e^{-j\omega}} + \pi\delta((\omega))_{2\pi}$
$\delta[n - n_0]$	$\Longleftrightarrow$	$e^{-j\omega n_0}$
$(n + 1)a^n u[n],\quad \|a\| < 1$	$\Longleftrightarrow$	$\dfrac{1}{(1 - ae^{-j\omega})^2}$
$\dfrac{(n + k - 1)!}{n!(k - 1)!}a^n u[n],\quad \|a\| < 1$	$\Longleftrightarrow$	$\dfrac{1}{(1 - ae^{-j\omega})^k}$

[a]The DTFT is always periodic in ω with period 2π. Thus impulses in the frequency domain are actually impulse trains with a period of 2π. This is indicated by the notation $((\cdot))_{2\pi}$. $\delta(\omega)$ is a Dirac delta function.

The Fourier transform of a continuous-time signal is defined by the integral

$$X_a(\Omega) = \int_{-\infty}^{\infty} x_a(t)e^{-j\Omega t}dt. \tag{3.3}$$

A number of similarities and differences between equations (3.1) and (3.3) become evident when they are compared. For example, both are complex functions, and thus are displayed in terms of real and imaginary parts or in terms of magnitude and phase. Unlike the Fourier transform, however, the DTFT is periodic in frequency with period 2π. When $x[n]$ is of finite length, the DTFT can be evaluated numerically for selected frequencies, but the Fourier transform cannot, in general. The units of the frequency variables are also different: ω has units of radians while Ω is measured in radians per second.

The DTFT is used frequently because many signals, systems, and signal processing operations can be described more conveniently in the frequency domain than in the time domain. A number of properties of the DTFT are listed in Table 3.2; these can be very helpful in making difficult problems easy to solve.

Table 3.2. Properties of the Discrete-Time Fourier Transform[a]

$x[n]$	$\Longleftrightarrow$	$X(e^{j\omega})$
$x[n - n_0]$	$\Longleftrightarrow$	$e^{-j\omega n_0} X(e^{j\omega})$
$e^{j\omega_0 n} x[n]$	$\Longleftrightarrow$	$X(e^{j(\omega - \omega_0)})$
$x^*[n]$	$\Longleftrightarrow$	$X^*(e^{-j\omega})$
$x[-n]$	$\Longleftrightarrow$	$X(e^{-j\omega})$
$x[n] * y[n]$	$\Longleftrightarrow$	$X(e^{j\omega})Y(e^{j\omega})$
$x[n]y[n]$	$\Longleftrightarrow$	$\dfrac{1}{2\pi} \displaystyle\int_{-\pi}^{\pi} X(e^{j\theta})Y(e^{j(\omega - \theta)})\, d\theta$
$x[n] - x[n - 1]$	$\Longleftrightarrow$	$(1 - e^{-j\omega})X(e^{j\omega})$
$\displaystyle\sum_{k=-\infty}^{n} x[k]$	$\Longleftrightarrow$	$\dfrac{1}{1 - e^{-j\omega}}X(e^{j\omega}) + \pi X(e^{j0})\delta((\omega))_{2\pi}$
$nx[n]$	$\Longleftrightarrow$	$j\dfrac{dX(e^{j\omega})}{d\omega}$
$\Re\{x_e[n]\}$	$\Longleftrightarrow$	$ev\{\Re(X(e^{j\omega}))\}$
$j\,\Im m\{x_e[n]\}$	$\Longleftrightarrow$	$jev\{\Im m(X(e^{j\omega}))\}$
$\Re\{x_0[n]\}$	$\Longleftrightarrow$	$jod\{\Im m(X(e^{j\omega}))\}$
$j\,\Im m\{x_0[n]\}$	$\Longleftrightarrow$	$od\{\Re(X(e^{j\omega}))\}$

[a]The symbol $\delta((\omega))_{2\pi}$ is the Dirac delta function. The notation $((\cdot))_{2\pi}$ reflects the fact that the DTFT is always periodic in ω with period 2π.

Two additional properties relate measurements in the time domain to measurements in the frequency domain. *Parseval's relation* relates signal energy in the time domain to measurements made from the DTFT.

$$\sum_{n=-\infty}^{\infty} |x[n]|^2 = \frac{1}{2\pi} \int_{-\pi}^{\pi} |X(e^{j\omega})|^2 \, d\omega.$$

The *initial value relations*

$$x[0] = \frac{1}{2\pi} \int_{-\pi}^{\pi} X(e^{j\omega}) \, d\omega$$

$$X(e^{j0}) = \sum_{n=-\infty}^{\infty} x[n]$$

are frequently useful for checking computations.

The exercises in this section examine the frequency domain and the ways in which the discrete-time Fourier transform may be used as an analysis tool.

EXERCISE 3.1.1. **Response of a System to a Complex Exponential**

An LTI system is completely characterized by its impulse response. Here the output of an LTI system is examined when the input is a complex exponential. Consider the system with impulse response $h[n]$ given by

$$h[n] = 0.03\delta[n] + 0.4\delta[n-1] + 0.54\delta[n-2] + 0.2\delta[n-3] - 0.2\delta[n-4].$$

Create a signal file containing this impulse response using the *create file* option in **x siggen**. Generate a set of six 128-point complex exponentials of the form $e^{j\omega_k n}$ with the following discrete-time frequencies: $\omega_1 = \pi/6$, $\omega_2 = \pi/3$, $\omega_3 = \pi/2$, $\omega_4 = 2\pi/3$, $\omega_5 = 5\pi/6$, and $\omega_6 = 8\pi/9$. This may be done easily by using the *exponential $Ke^{j\alpha n}$* option in **x siggen**. The parameters for α should be complex in this case, e.g., $0 \ \pi/2$, $0 \ \pi/3$, etc. Display the real and imaginary parts of the complex exponential and observe that they are sinusoids with unity amplitude. Convolve each complex exponential with the system impulse response, $h[n]$, to form $y_1[n]$, $y_2[n]$, ..., $y_6[n]$.

(a) Examine and record the magnitude of each of the output sequences in the steady state region, i.e., $4 \leq n \leq 128$. The function **x view2** can be used to examine the magnitudes of these outputs one above the other. You will observe that the outputs are essentially constant. The only differences are that the magnitudes of the output sequences may differ from those of the inputs and that the output sequences may have been shifted slightly in time.

(b) Construct a graph with amplitude along the vertical axis and frequency in the range $(0 \leq \omega \leq \pi)$ along the horizontal axis. Plot the peak magnitude values for $y_1[n]$, $y_2[n]$, ..., $y_6[n]$ on the graph—one point for each of the six frequencies.

Connect the points to form a continuous plot. Your result should approximate the discrete-time Fourier transform magnitude, $|H(e^{j\omega})|$.

(c) The discrete-time Fourier transform, or DTFT, is a frequency-domain representation that is most often displayed in the frequency range $-\pi \leq \omega < \pi$. Plots of the DTFT are continuous curves. The computer is actually displaying discrete sample values of the DTFT and connecting these points with straight lines. Use the **x dtft** function to display the DTFT magnitude, $|H(e^{j\omega})|$. Sketch the transform magnitude in the range $0 \leq \omega \leq \pi$. How does it compare to your plot in part (b)?

The discrete-time Fourier transform of a signal or system can be interpreted as its response to an infinite set of complex sinusoids that collectively span the frequency range.

EXERCISE 3.1.2. **Symmetry in the Fourier Transform**

In Chapter 2 the concept of even and odd symmetry was introduced. It was shown that any real sequence, $x[n]$, could be expressed as a sum of its even and odd parts where the even part was defined as

$$ev\{x[n]\} = (x[n] + x[-n])/2$$

and the odd part of $x[n]$ was defined as

$$od\{x[n]\} = (x[n] - x[-n])/2.$$

This concept was also extended so that any arbitrary complex sequence, $\tilde{x}[n]$, could be expressed in terms of the even and odd parts of its real and imaginary components, i.e.,

$$\tilde{x}[n] = \underbrace{a_e[n]}_{\text{real}-\text{even}} + \underbrace{a_o[n]}_{\text{real}-\text{odd}} + j \underbrace{b_e[n]}_{\text{imag}-\text{even}} + j \underbrace{b_o[n]}_{\text{imag}-\text{odd}}.$$

The DTFT can be similarly expressed in terms of these components,

$$\tilde{X}(e^{j\omega}) = \underbrace{C_e(e^{j\omega})}_{\text{real}-\text{even}} + \underbrace{C_o(e^{j\omega})}_{\text{real}-\text{odd}} + j \underbrace{D_e(e^{j\omega})}_{\text{imag}-\text{even}} + j \underbrace{D_o(e^{j\omega})}_{\text{imag}-\text{odd}}.$$

In this exercise the even and odd symmetry properties with respect to the discrete-time Fourier transform are investigated.

(a) Generate an arbitrary 5-point complex sequence, $\tilde{x}[n]$. Do this by using **x rgen** to create two random 5-point real sequences. Use **x gain** with the *gain* parameter equal to "0 1" (which represents the complex number j) to transform one of the real sequences into a purely imaginary signal, then add the sequences together using **x add**. Sketch the real and imaginary parts of $\tilde{x}[n]$.

(b) Using the computer, decompose $\tilde{x}[n]$ into its four components, $a_e[n]$, $a_o[n]$, $b_e[n]$, $b_o[n]$, and provide a sketch for each. The functions **x realpart** and **x imagpart** can be used to extract the real and imaginary parts, respectively. The even and odd components can be obtained by using the **x reverse**, **x add**, **x subtract**, and **x gain** functions. Next compute the DTFT of $\tilde{x}[n]$. Decompose $\tilde{X}(e^{j\omega})$ into its components $C_e(e^{j\omega})$, $C_o(e^{j\omega})$, $D_e(e^{j\omega})$, and $D_o(e^{j\omega})$ in the same way. Recall from Chapter 1 that the function **x dtft** generates a 513 point file "_dtft_" that contains samples of the transform. (Note that this file name begins and ends with underbars.) You may reverse, add, and process this file appropriately to generate $C_e(e^{j\omega})$, $C_o(e^{j\omega})$, $D_e(e^{j\omega})$, and $D_o(e^{j\omega})$.

(c) Now individually take the discrete-time Fourier transform of $a_e[n]$, $a_o[n]$, $jb_e[n]$, and $jb_o[n]$. Summarize your findings by matching the time-domain and frequency-domain components listed below.

Time Domain	Frequency Domain
$a_e[n]$	$C_e(e^{j\omega})$
$a_o[n]$	$C_o(e^{j\omega})$
$jb_e[n]$	$jD_e(e^{j\omega})$
$jb_o[n]$	$jD_o(e^{j\omega})$

These symmetry properties that you have observed apply for any arbitrary complex sequence.

EXERCISE 3.1.3. Proving the Symmetry Properties

In Chapter 2 and in the previous exercise the concept of a decomposition into even and odd signal components was discussed. You saw that any arbitrary signal (real or complex) could be expressed in terms of real-even, real-odd, imaginary-even, and imaginary-odd components. In the previous exercise you were shown experimentally that each of these components transforms to one of four related frequency-domain components.

Consider the definitions for the even and odd components in the time domain and the frequency domain:

$$a_e[n] = \tfrac{1}{2}(a[n] + a[-n])$$

$$a_o[n] = \tfrac{1}{2}(a[n] - a[-n])$$

$$C_e(e^{j\omega}) = \tfrac{1}{2}(C(e^{j\omega}) + C(e^{-j\omega}))$$

$$C_o(e^{j\omega}) = \tfrac{1}{2}(C(e^{j\omega}) - C(e^{-j\omega}))$$

where $a[n]$ and $C(e^{j\omega})$ are assumed to be purely real signals. Using the definition of the DTFT, prove analytically that the relationships you found in the previous exercise are always true. (*Hint:* You may wish to use the fact that

$$\sum_{n=-\infty}^{\infty} f[n] = 0$$

if $f[n]$ is odd. You may also wish to determine whether each of the components is either even or odd. Then determine if each is real or imaginary.)

EXERCISE 3.1.4. Hermitian Symmetry

The two previous exercises established the relationships between the real, imaginary, even, and odd parts of signals in the time and frequency domains. A special case of this general result occurs when the time signal, $x[n]$, is purely real. In this case, the real part of the DTFT is an even function and the imaginary part is an odd function. This condition is known as *Hermitian symmetry.*

(a) Using the function **x dtft** sketch the real and imaginary parts of the DTFT of the signal $cc[n]$ contained in the file **cc**. Observe the symmetry in the real and imaginary parts. What type of symmetry is present in the magnitude and phase of the DTFT of this signal, i.e., which is an even function and which is odd?

(b) Now use **x dtft** to examine the DTFT of the *complex* signal, $dd[n]$ contained in the file **dd**. Sketch the real and imaginary parts of its DTFT. Also sketch the magnitude and phase.

(c) It is very common to display the DTFT only over the range from $0 \le \omega \le \pi$ due to Hermitian symmetry. Based on your observations in part (b), when would it be inappropriate to display the transform in this limited range?

EXERCISE 3.1.5. Periodicity of the Discrete-Time Fourier Transform

The discrete-time Fourier transform shares many characteristics with the continuous-time Fourier transform. A major difference, however, is that the DTFT is periodic in ω with period 2π.

To examine the issue of periodicity, consider the definition

$$X(e^{j\omega}) = \sum_{n=-\infty}^{\infty} x[n]e^{-j\omega n}$$
$$= \cdots + x[-1]e^{j\omega} + x[0] + x[1]e^{-j\omega} + x[2]e^{-j2\omega} + \cdots.$$

Carefully examine each term in the DTFT and observe that it repeats in frequency every 2π radians. This is because $e^{-j2\pi n} = 1$ for $n = 0, \pm 1, \pm 2, \ldots$.

Use **x dtft** to compute the DTFT of the signal $aa[n]$ contained in the file **aa**. By examining the **x dtft** display, estimate the numerical values for each of the following using the periodicity property:

(a) $|AA(e^{j\pi/4})|$.

(b) $|AA(e^{j3\pi/4})|$.

(c) $|AA(e^{j9\pi/4})|$.

(d) $|AA(e^{j17\pi/4})|$.

Note that **x dtft** only displays the DTFT in the range between $-\pi$ and π.

EXERCISE 3.1.6. **The Shift Property**

The shift property of the DTFT says that shifting a signal in time by n_0 causes its DTFT to be multiplied or modulated by the complex exponential $e^{-j\omega n_0}$. In other words, if $x[n]$ has a DTFT $X(e^{j\omega})$, then $x[n - n_0]$ has DTFT $X(e^{j\omega})e^{-j\omega n_0}$.

(a) Display and sketch the 15-point signal $ee[n]$ contained in the file **ee** using **x view**. Using **x dtft**, display and sketch the imaginary part. Observe that since $ee[n]$ is an odd function, the DTFT is purely imaginary (i.e., its real part is zero), and consequently only one sketch is necessary for its complete representation. Practically speaking, the real part of the DTFT shown in the plot is only approximately zero due to limited numerical precision in the computation.

(b) Consider the signals

$$y_0[n] = ee[n - 1]$$

and

$$y_1[n] = ee[n + 1].$$

Using the **x lshift** and **x dtft** functions, sketch their frequency responses. You will observe that the resulting frequency response is no longer purely imaginary.

Euler's formula states that

$$e^{\pm j\omega n_0} = \cos \omega n_0 \pm j \sin \omega n_0.$$

Therefore, the shifted signals will have the form

$$Y_i(e^{j\omega}) = EE(e^{j\omega}) \cos \omega n_0 \pm j EE(e^{j\omega}) \sin \omega n_0. \tag{3.4}$$

Display and sketch the real and imaginary parts of $Y_0(e^{j\omega})$ and $Y_1(e^{j\omega})$ and observe the frequency-domain effects of shifting in time.

(c) By changing the summation variable, n in the DTFT definition (i.e., letting $n \rightarrow n - n_0$), derive the shift property.

EXERCISE 3.1.7. **The Modulation Property**

The modulation property is the dual of the shift property. It states that multiplying a sequence, $x[n]$, by a complex exponential $e^{j\alpha n}$ shifts its DTFT in frequency by α. If $x[n]$ has the DTFT $X(e^{j\omega})$, then $x[n]e^{j\alpha n}$ has the DTFT $X(e^{j(\omega - \alpha)})$. Consider the signal $ee[n]$ contained in the file **ee**. Modulate this signal by $e^{j\pi n/4}$ using **x cexp**.

(a) Display and sketch $|EE(e^{j\omega})|$ the DTFT magnitude of $ee[n]$, and also the DTFT magnitude of $e^{j\pi n/4}ee[n]$.

(b) Analytically derive the modulation property by explicitly taking the DTFT of $x[n]e^{j\alpha n}$.

EXERCISE 3.1.8. **The Convolution Property**

The convolution property is one of the most important and useful properties of the DTFT. It states that when two signals are convolved, their respective DTFTs are multiplied. In other words, the DTFT of $x[n] * h[n]$ is $X(e^{j\omega})H(e^{j\omega})$. Consider the sequences $ee[n]$ and $ff[n]$ contained in the files **ee** and **ff**.

(a) Display and sketch the magnitude responses of $EE(e^{j\omega})$ and $FF(e^{j\omega})$. Based on these plots, sketch the magnitude of $EE(e^{j\omega})\ FF(e^{j\omega})$.

(b) Using the **x convolve** function, compute

$$y[n] = ee[n] * ff[n].$$

Display and sketch the DTFT magnitude of $y[n]$.

(c) Consider deriving the convolution property. Recall that the convolution of two signals $x[n]$ and $h[n]$ is defined by the convolution sum

$$x[n] * h[n] = \sum_{m=-\infty}^{\infty} x[m]h[n - m].$$

Apply the DTFT definition to the convolution sum. Observe that this double summation may be rewritten as

$$\sum_{m=-\infty}^{\infty} x[m] \underbrace{\sum_{n=-\infty}^{\infty} h[n - m]e^{-j\omega n}}_{\text{DTFT of } h[n-m]}$$

since $x[m]$ has no dependence on the variable n. Using the definition of the DTFT and the shift property, show that this expression reduces to $X(e^{j\omega})H(e^{j\omega})$.

EXERCISE 3.1.9. **The Multiplication Property**

The multiplication and convolution properties are duals in the sense that multiplication in the time domain implies convolution in the frequency domain and convolution in the time domain implies multiplication in the frequency domain. The DTFT of a product of two sequences is the convolution of their individual DTFTs, i.e.,

$$\text{DTFT}\{x[n]h[n]\} = \frac{1}{2\pi}X(e^{j\omega}) \circledast H(e^{j\omega})$$

$$= \frac{1}{2\pi} \int_{-\pi}^{\pi} X(e^{j\theta})H(e^{j(\omega-\theta)})\, d\theta \qquad (3.5)$$

where the circled star is used to denote a form of convolution called *circular* or *periodic* convolution.

(a) Taking the DTFT of $x[n]h[n]$ results in

$$\text{DTFT}\{x[n]h[n]\} = \sum_{n=-\infty}^{\infty} x[n]h[n]e^{-j\omega n}.$$

By expressing $x[n]$ as the inverse DTFT of $X(e^{j\omega})$ and substituting, this equation becomes

$$\text{DTFT}\{x[n]h[n]\} = \sum_{n=-\infty}^{\infty} \underbrace{\frac{1}{2\pi} \int_{-\pi}^{\pi} X(e^{j\omega})e^{j\theta n} \, d\theta}_{x[n]} \, h[n]e^{-j\omega n}.$$

By interchanging the order of the sum and integration operations, show analytically that the multiplication property is valid.

(b) Circular convolution is a topic that will be considered in more detail later in this text. For the present we will consider only a simple example. Display and sketch the DTFT magnitudes of the sequences $ff[n]$ and $gg[n]$ contained in files **ff** and **gg**. The circular convolution of $FF(e^{j\omega})$ and $GG(e^{j\omega})$ is very much like their linear convolution except that the two signals are periodic and the integration is confined to one period. Display and sketch the magnitude of

$$FF(e^{j\omega}) \circledast GG(e^{j\omega})$$

using the multiplication property defined in equation (3.5), i.e., use the functions **x multiply** and **x dtft**.

EXERCISE 3.1.10. **The Initial Value Theorem**

The initial value theorem states that

$$x[0] = \frac{1}{2\pi} \int_{-\pi}^{\pi} X(e^{j\omega}) \, d\omega,$$

which is proportional to the area under a graph of the DTFT. The analogous relationship with the domains reversed is

$$X(e^{j0}) = \sum_{n=-\infty}^{\infty} x[n].$$

These properties can be very useful in solving certain problems.

(a) Write the definition of the inverse DTFT. Let n equal zero and show that the result reduces to the initial value theorem.

(b) Now use the definition of the forward DTFT to derive the other initial value relationship.

(c) Using the **x lshift** function, create the signal $vv[n] = gg[n + 5]$, where $gg[n]$ is the signal contained in the file **gg**. Now use **x dtft** to display the real and imaginary parts. Sketch the real part of $VV(e^{j\omega})$ and estimate the area under the curve. Since **x dtft** produces an output file _dtft_ containing the 513 samples of the DTFT used in the graphics, the function **x summer** can be used to sum the frequency domain samples. To obtain the area under the curve, you must multiply this sum by $2\pi/513$. Now examine $vv[n]$ using **x view** and record the value of $vv[0]$. Verify that

$$vv[0] = \frac{1}{2\pi}(\text{area under } VV(e^{j\omega})).$$

EXERCISE 3.1.11. **Some Fourier Transform Pairs**

A number of signals arise frequently in digital signal processing. Remembering their discrete-time Fourier transforms can be very useful. This exercise will add to the list of familiar discrete-time Fourier transform pairs.

(a) Compute and sketch the DTFT magnitudes of the following finite length sequences using **x siggen, x multiply**, and **x dtft**:

   (i) $\sin(\frac{\pi n}{8})(u[n] - u[n - 64])$.

   (ii) $\cos(\frac{\pi n}{8})(u[n] - u[n - 64])$.

   (iii) $u[n] - u[n - 32]$.

   (iv) $\frac{\sin(\pi n/2)}{\pi n}(u[n + 20] - u[n - 21])$. You may use **x fdesign** with the rectangular window option to create $\sin[(\pi n/2)/\pi n]$.

   (v)   A 64-point square wave with period 8.

   (vi) A 64-point impulse train with period 8.

Notice that the discrete-time Fourier transforms of these signals are slightly different from the Fourier transforms of the related continuous-time signals. Explain the differences.

(b) It can be shown analytically that the discrete-time Fourier transform of a sine wave is a pair of impulses; however, the results that you observed on the computer were not impulses. Explain this apparent contradiction.

3.2 FILTERS

The term "filter" describes a very large variety of different systems. However, it is commonly used to refer to LTI systems with frequency-selective frequency responses. These filters allow certain regions of the spectrum to remain undisturbed while other spectral regions are attenuated. Familiar examples include lowpass, highpass, and bandpass filters.

Filters are specified by their frequency-domain magnitude characteristics. Ideally, the frequency response of a filter should display only two types of behavior: it should have one or more spectral regions of constant nonzero amplitude (called the *passband(s)*) and one or more regions where the frequency response is zero (called the *stopband(s)*). These ideal filters are not realizable in practice. Realizable filters will also contain intermediate regions called *transition bands*. Because the frequency response must be continuous, a transition band must lie between each passband and stopband.

Realizable filters are designed to approximate ideal filters to within certain prescribed tolerances. These tolerances are defined by the widths of the transition regions, the maximum passband deviation, and the maximum stopband deviation. These parameters are illustrated in Fig. 3.1 for a lowpass filter. The effect is to define a region on the graph in which the magnitude response of the filter must lie. The ideal frequency response is shown by the solid black line. The following parameters define the match between the digital filter and the ideal:

δ_p passband ripple,

δ_s stopband ripple,

ω_ℓ lower cutoff frequency,

ω_u upper cutoff frequency,

ω_c nominal cutoff frequency.

For high-quality filters that closely approximate the ideal response, the ripples in the stopband may not be visible in a plot of the magnitude response. In these cases, the log magnitude response, defined as $20\log_{10}|H(e^{j\omega})|$, may be more useful. The vertical axis of the log magnitude plot is measured in decibels and clearly shows the variation in the stopband region. Thus, an alternate way of specifying the maximum stopband deviation is through the *attenuation* where

$$\text{attenuation} \stackrel{\triangle}{=} -20\log_{10}\delta_s \quad (\text{dB}).$$

The *phase response* of the filter is defined as

$$\angle H(e^{j\omega}) = \tan^{-1}\frac{\Im m\{H(e^{j\omega})\}}{\Re e\{H(e^{j\omega})\}} + \frac{\pi}{2}(1 + \text{sgn}(\Re e\{H(e^{j\omega})\})).$$

Since the phase function is a multivalued function, the principal value is generally displayed. This results in a function with an amplitude that ranges between $-\pi$ and $+\pi$. The phase response of a filter can be important in certain applications, although historically greater attention has been given to the magnitude response.

Digital filters can be divided into two general categories based on their lengths: finite impulse response (FIR) filters and infinite impulse response (IIR) filters. FIR filters have impulse responses that are finite in duration; they can be implemented conveniently using convolution. IIR filters, on the other hand, are infinite in duration and are implemented using difference equations. Much can be said about digital filters, their properties, and methods for their design, but this is the topic of the companion text, *Digital Filters: A Computer Laboratory Text*. The few exercises that follow try to provide some limited familiarity with this topic.

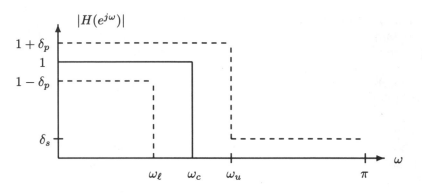

Figure 3.1. Tolerances for specifying the acceptable performance of a lowpass filter. The ideal filter $H(e^{j\omega})$ is shown by the solid line.

EXERCISE 3.2.1. Digital Filters

When the frequency response of a linear time-invariant system allows a certain region of the spectrum to be passed undisturbed and rejects or nullifies other frequency regions, the system is often called a filter. Consider the two FIR filters $h_1[n]$ and $h_2[n]$ with impulse responses

$$h_1[n] = 0.036\delta[n] - 0.036\delta[n-1] - 0.29\delta[n-2] + 0.56\delta[n-3]$$
$$-0.29\delta[n-4] - 0.036\delta[n-5] + 0.036\delta[n-6]$$
$$h_2[n] = -0.09\delta[n] + 0.12\delta[n-1] + 0.5\delta[n-2]$$
$$+0.5\delta[n-3] + 0.12\delta[n-4] - 0.09\delta[n-5].$$

In addition, consider the IIR filter that is described by the difference equation

$$y[n] - 0.57y[n-1] + 0.88y[n-2] - 0.26y[n-3] + 0.09y[n-4]$$
$$= 0.11x[n] + 0.27x[n-1] + 0.37x[n-2] + 0.27x[n-3] + 0.11x[n-4].$$

Use the *create file* option of **x siggen** to create files for each of these filters. Display and sketch the magnitude responses for each filter using **x dtft**. Classify each as a lowpass or highpass. What are the approximate cutoff frequencies and transition widths in radians for each of these filters? What is the stopband attenuation (in dB) for each? Estimate these values by inspecting the DTFT plots.

EXERCISE 3.2.2. Time-Domain Filter Characteristics

In the preceding exercise, you examined the frequency characteristics of some LTI filters. In this exercise, you will investigate time-domain characteristics of ideal lowpass filters.

The impulse response, $h[n]$, of an ideal lowpass filter can be expressed analytically as

$$h[n] = \frac{\sin \omega_0 n}{\pi n}.$$

Give a rough sketch of this signal for $\omega_0 = \pi/2$, $\pi/4$, $\pi/6$, and $\pi/8$. The function **x fdesign** (using the rectangular window option) can be used to construct lowpass filters that have this time domain characteristic.

(a) Design four 32-point lowpass filters with cutoff frequencies as specified above. Examine them carefully in both the time domain and the frequency domain. In the time domain each impulse response has a peaked cluster of samples centered in the middle of the sequence. This cluster is called the *main lobe*. How is the width of the main lobe in the impulse response related to the width of the pass-band of the filter?

(b) The step response of a filter is the output when the input is an ideal unit step, $u[n]$. In this exercise, approximate the step function by a block of length 100 created using **x siggen**. Sketch the step responses for the set of lowpass filters used in part (a). Truncate your result to 50 samples using **x truncate** to avoid displaying the erroneous points due to the block approximation of the unit step. Determine the relationship between the slope of the step response in the middle of the transition and the main lobe width of the impulse response. Observe that the ripple heights (overshoot and undershoot) in the step response are not significantly affected by the nominal cutoff frequency.

EXERCISE 3.2.3. Filter Design

There are many programs available for designing FIR filters. This text does not address filter design procedures. However, the use of filters and the ability to design them are important if complex systems are to be constructed. The program **x fde-sign** is a filter design program based on the *window design procedure*. Three choices for the window can be chosen: a rectangular window, a Hanning (Von Hann) window, or a Hamming window. Design a 32-point and a 64-point lowpass filter with cutoff $\pi/3$ using each of these window options. Measure the attenuation and transition band in each case using **x dtft**.

(a) Which window gives the greatest attenuation? What is the value of that attenuation? Does it seem to depend on the filter length?

(b) Which option gives the narrowest transition band for a given filter length? How does the transition band depend on the filter length?

EXERCISE 3.2.4. Filtering a Sinusoid and a Random Sequence

The frequency domain provides useful insight in understanding filtering. Create a file containing the filter $h_1[n]$ defined in Exercise 3.2.1 using the *create file* option of **x siggen**.

(a) Display and sketch the DTFT magnitude of this filter.

(b) Create the following two signals using **x siggen**, **x add**, and **x rgen**.

    (i)   The first 256 points of

$$x_1[n] = \sum_{k=1}^{4} \sin \frac{\pi k}{5} n.$$

(ii)   A 256-point random signal $x_2[n]$.

Display and sketch the DTFT magnitudes of $x_1[n]$ and $x_2[n]$.

(c)   Consider $x_1[n]$ and $x_2[n]$ as inputs to the filter with impulse response $h_1[n]$. Without the aid of the computer, sketch the DTFT magnitude of the outputs. Verify your results using the computer.

An LTI system can be used to alter the spectral shape of the input. For the random signals generated by **x rgen**, the DTFT has energy that is uniformly distributed in the frequency domain. Thus the frequency response of the LTI system shapes the spectrum of the random signal.

3.3   LINEAR PHASE

The DTFT is a complex function that can be expressed in terms of either its real and imaginary parts or its magnitude and phase—the latter being more common. An important subclass of signals is the class for which the DTFT phase is linear. Real, linear phase signals have the property that their discrete-time Fourier transforms can be written as $A(e^{j\omega})e^{-j\omega n_0}$ where $A(e^{j\omega})$ is either purely real or purely imaginary. In this context, $A(e^{j\omega})$ plays the role of an amplitude function and $-\omega n_0$ plays the role of the phase. Since this phase component is a linear function of ω, signals of this form are termed *linear-phase sequences*.

The linear-phase condition imposes a well-defined symmetric structure on the signal in the time domain. For odd length real sequences, the samples are either symmetric or antisymmetric about a midpoint as shown in Fig. 3.2. For even length sequences, the symmetry characteristics are similar, except that the point of symmetry lies midway between two sample values. For finite length sequences that begin at N_1 and end at N_2 ($N_1 < N_2$), the midpoint, n_0 is at

$$n_0 = \frac{N_2 + N_1}{2}.$$

When the sequence length is odd, n_0 is an integer, but when the sequence length is even, n_0 is a half integer.

An interesting subset of linear-phase sequences is the class of odd length sequences with a midpoint at $n = 0$. These are called *zero-phase* sequences because n_0 is zero and the phase term is also identically zero.

EXERCISE 3.3.1.   **Zero Phase**

Zero-phase sequences are odd length linear-phase sequences that are centered at the origin. Due to the sequence symmetry, the discrete-time Fourier transform of a

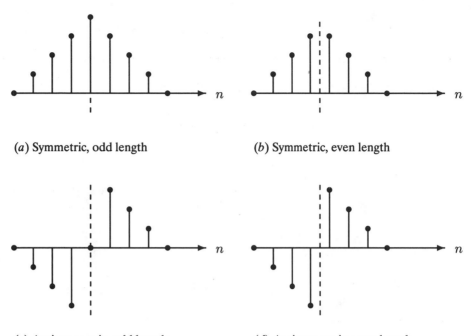

(*a*) Symmetric, odd length   (*b*) Symmetric, even length

(*c*) Antisymmetric, odd length   (*d*) Antisymmetric, even length

Figure 3.2.   The four types of linear-phase sequences.

real odd length, symmetric (even function) zero-phase sequence $h[n]$ can always be expressed as

$$H(e^{j\omega}) - \sum_{n=-L}^{L} h[n]e^{-j\omega n} - h[0] + 2\sum_{n=1}^{L} h[n]\cos\omega n. \qquad (3.6)$$

Create an arbitrary 13-point zero-phase sequence, $e[n]$. There are many ways to do this, one of which is to use **x fdesign**. Sketch the sequence and its DTFT. Now create

$$f[n] = e[n] + \delta[n]$$

and sketch $F(e^{j\omega})$. Compare $E(e^{j\omega})$ and $F(e^{j\omega})$ and explain why they have similar spectral shapes.

EXERCISE 3.3.2. A Cascaded System

The frequency response of an LTI system provides useful information about its behavior. Consider the LTI systems with impulse responses

$$h_1[n] = -0.04\delta[n] + 0.04\delta[n-1] + 0.3\delta[n-2] - 0.6\delta[n-3]$$
$$+0.3\delta[n-4] + 0.04\delta[n-5] - 0.04\delta[n-6]$$

$$h_2[n] = 0.09\delta[n] - 0.12\delta[n-1] - 0.5\delta[n-2]$$
$$-0.5\delta[n-3] + 0.12\delta[n-4] + 0.09\delta[n-5].$$

Use the *create file* option of **x siggen** to create files for each filter.

(a) Using **x dtft** display and sketch the magnitude responses of $h_1[n]$ and $h_2[n]$.

(b) Now consider a new system composed of the cascade of $h_1[n]$ and $h_2[n]$. The impulse response of the new system, which we will call $h_3[n]$, can be shown to equal

$$h_3[n] = h_1[n] * h_2[n].$$

Without the aid of the computer, give a rough sketch of the magnitude response of $h_3[n]$. Verify your answer by explicitly computing $h_3[n]$ and taking its DTFT.

EXERCISE 3.3.3. A Parallel System

Consider two LTI systems defined by the impulse responses

$$h_1[n] = 0.04\delta[n] - 0.04\delta[n-1] - 0.3\delta[n-2] + 0.6\delta[n-3]$$
$$-0.3\delta[n-4] - 0.04\delta[n-5] + 0.04\delta[n-6]$$
$$h_2[n] = 0.09\delta[n] - 0.12\delta[n-1] - 0.5\delta[n-2]$$
$$-0.5\delta[n-3] + 0.12\delta[n-4] + 0.09\delta[n-5].$$

Use the *create file* option of **x siggen** to create files for these filters.

(a) Using **x dtft** display and sketch the magnitude responses for $h_1[n]$ and $h_2[n]$.

(b) Now consider a new system $h_3[n]$ composed of the sum of $h_1[n]$ and $h_2[n]$ where

$$h_3[n] = h_1[n] + h_2[n].$$

Without the aid of the computer, give a rough sketch of the magnitude response of $h_3[n]$. Verify your answer by explicitly computing $h_3[n]$ using **x add** and taking its DTFT magnitude.

Sampling

4

Most signals originate as continuous-time waveforms. Sampling, or analog-to-digital (A/D) conversion, allows these analog signals to be represented in digital form. Ideal sampling is defined by

$$x[n] = x_a(t)|_{t=nT} = x_a(nT), \qquad -\infty < n < \infty. \tag{4.1}$$

The continuous-time (analog) signal, $x_a(t)$, is a function of the continuous (time) variable t and $x[n]$ is the sequence of sample values. The constant T, known as the *sampling period*, defines the time spacing between samples. Its reciprocal, $f_s = 1/T$, is the *sampling rate,* which is measured in samples per second.[1]

This chapter addresses many of the issues that surround A/D conversion and its dual process, D/A conversion. D/A conversion reconstructs bandlimited continuous-time signals from their sample values. The next section examines the processes of sampling and signal reconstruction when there is no quantization of the sample values. In the nonideal world sampling represents only one of two aspects of the digitization process. In addition to the discretization of the t variable, the sample values must also be quantized in amplitude. Quantization is a nonlinear operation that will be treated separately in Section 4.2. The final section of this chapter looks at sampling-rate conversion. This is the process in which a sequence of sample values that have been taken at one sampling rate are digitally mapped into another sequence of sample values that are taken at a different rate. It is often important that the process of sampling-rate conversion preserve the essential properties of the Fourier transforms of all of the signals that are involved.

---

[1]The term *sampling frequency* is often used in lieu of *sampling rate*, in which case the unit of measurement is Hz or cycles per second.

4.1 SAMPLING AND RECONSTRUCTION

Any discrete-time representation for continuous-time waveforms must satisfy several requirements: it must be invertible, it must preserve the essential information in the waveform, and it must be implementable. Remarkably, periodic sampling meets all of these conditions when the continuous-time waveform is bandlimited. We can verify this by looking at sampling in both the time and frequency domains.

Let $x_a(t)$ be the continuous-time signal that is sampled and let $X_a(\Omega)$ be its continuous-time Fourier transform. These two functions are related by

$$X_a(\Omega) = \int_{-\infty}^{\infty} x_a(t)e^{-j\Omega t}dt$$

where $X_a(\Omega)$ is a complex function of the frequency variable Ω with units of radians per second (rad/sec), and the subscript "a" signifies an analog or continuous-time waveform. An analogous Fourier relationship exists for digital signals as discussed in Chapter 3. The discrete-time sequence of samples, $x[n]$, and its discrete-time Fourier transform, $X(e^{j\omega})$, are related by

$$X(e^{j\omega}) = \sum_{n=-\infty}^{\infty} x[n]e^{-j\omega n}$$

where $X(e^{j\omega})$ is a complex, continuous, and periodic function of the discrete-domain frequency variable ω, which is measured in radians. The ideal A/D converter receives as an input a continuous-time signal and produces the discrete-time signal $x[n]$ as its output, as shown in Fig. 4.1.

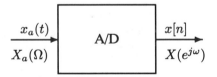

Figure 4.1.   An ideal A/D converter.

The analog and digital signals are related by equation (4.1) in the time domain. Their Fourier transforms are related by

$$X(e^{j\omega}) = \frac{1}{T} \sum_{r=-\infty}^{\infty} X_a\left(\Omega + \frac{2\pi r}{T}\right) \qquad \text{where } \Omega = \frac{\omega}{T}. \qquad (4.2)$$

This relationship states that $X(e^{j\omega})$ is a scaled superposition of shifted copies of $X_a(\Omega)$ that have been normalized in frequency. A graphical interpretation of this formula is given in Fig. 4.2. The spectrum shown at the top of Fig. 4.2 shows a signal that is confined to the frequency region between $-\Omega_0$ and Ω_0. Such signals are said to be *bandlimited* to Ω_0. The middle graph shows the resulting DTFT when $\Omega_0 < \pi/T$.

The bottom graph shows the same DTFT for the case where $\Omega_0 > \pi/T$. Equation (4.2) states that the DTFT is composed of a sum of an infinite number of identical, uniformly shifted replicas of $X_a(\Omega)$. This sum produces the periodic signal $X(e^{j\omega})$. The spacing between the midpoints of the spectral copies is 2π radians. This is consistent with the fact that the period of any DTFT is 2π radians.

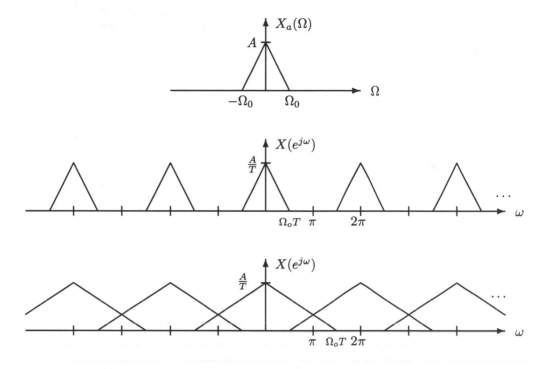

Figure 4.2.   A look at sampling from the frequency domain. (Top) Spectrum of a bandlimited continuous-time signal. (Middle) Spectrum of the sampled signal when $\Omega_0 < \pi/T$. (Bottom) Spectrum of the sampled signal when $\Omega_0 > \pi/T$.

The dual process in which discrete signals are converted into continuous-time waveforms is called *digital-to-analog (or D/A) conversion*. A D/A converter implements the relation

$$y_a(t) = \sum_{n=-\infty}^{\infty} y[n]h_a(t-nT)$$

$$= \underbrace{\left\{ \sum_{n=-\infty}^{\infty} y[n]\delta(t-nT) \right\}}_{y_c(t)} *h_a(t)$$

$$= y_c(t) * h_a(t). \tag{4.3}$$

Using the bottom equation, signal reconstruction can be viewed as a two-step process. The sequence is first converted into a weighted impulse train, $y_c(t)$. This impulse train is then passed through a continuous-time filter with the impulse response $h_a(t)$. This is shown in Fig. 4.3. In the frequency domain equation (4.3) becomes

$$Y_a(\Omega) = Y_c(\Omega)H_a(\Omega), \tag{4.4}$$

where

$$Y_c(\Omega) = \mathcal{F}\left\{\sum_{n=-\infty}^{\infty} y[n]\delta(t - nT)\right\}$$

$$= \sum_{n=-\infty}^{\infty} y[n]e^{-j\Omega T n}$$

$$= Y(e^{j\Omega T}) \tag{4.5}$$

and the operator $\mathcal{F}\{\cdot\}$ denotes the (continuous-time) Fourier transform. The Fourier transform of the continuous-time impulse train $y_c(t)$ is identical to the DTFT of the sequence of samples except for the denormalization of the frequency variable. The original (analog) and normalized (discrete) frequency variables are related by

$$\omega = \Omega T. \tag{4.6}$$

The ideal D/A conversion process can be clearly seen in the frequency domain. The starting point is $Y(e^{j\omega})$, which is periodic in ω with period 2π, as shown in Fig. 4.4. The first step is to convert $y[n]$ into the weighted analog impulse train $y_c(t)$. In the frequency domain this corresponds to a change in the frequency variable: $\omega \Rightarrow \Omega T$. $Y_c(\Omega)$ is periodic in Ω with a period of $2\pi/T$. It contains an infinite number of periodic replicas of $Y_a(\Omega)$. The second and final step removes these replicas by filtering $Y_c(\Omega)$ with an ideal lowpass filter $H_a(\Omega)$ with a cutoff frequency of π/T and a gain of T.

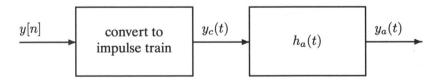

Figure 4.3.   An interpretation of the D/A conversion process.

A/D and D/A converters allow for discrete-time processing of continuous-time signals and are often used together. Assume that an ideal A/D converter is cascaded with an ideal D/A, as shown in Fig. 4.5. If the spectrum of $x_a(t)$ resembles the one shown in the top portion of Fig. 4.2 with $\Omega_0 < \pi/T$, the spectrum at the input to the filter in the D/A converter is the same as the one shown in the middle of Fig. 4.2. The output of the filter is equal to $x_a(t)$ when $h_a(t)$ is an ideal lowpass with gain T and cutoff frequency π/T radians. Therefore,

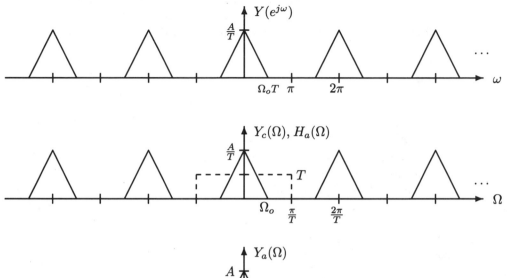

Figure 4.4.   Relationships between the spectra involved in ideal D/A conversion: (top) $Y(e^{j\omega})$; (middle) $Y_c(\Omega)$; (bottom) $Y_a(\Omega)$.

$$h_a(t) = T\frac{\sin \pi t/T}{\pi t} \tag{4.7}$$

(which is the impulse response of an ideal lowpass filter) and

$$y_a(t) = T\sum_{k=-\infty}^{\infty} y[k]\frac{\sin(\pi/T)(t - kT)}{\pi(t - kT)}. \tag{4.8}$$

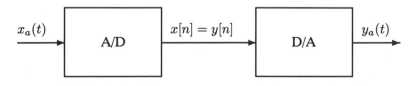

Figure 4.5.   A cascade of an ideal A/D converter with an ideal D/A.

The ideal lowpass filter in equation (4.7) is not causal and it cannot be realized electronically. Therefore, practical D/A converters make use of causal substitutes for it. Nevertheless, the ideal D/A converter is important because it represents the standard against which all practical D/A converters should be compared.

It should be clear that the ideal D/A converter inverts the sampling operation only when:

1. $X_a(\Omega)$ is bandlimited to Ω_0. By definition, an analog signal, $x_a(t)$, is *bandlimited* if

$$X_a(\Omega) = 0 \qquad |\Omega| > \Omega_0.$$

The frequency Ω_0 is called the *Nyquist frequency*. An example of a bandlimited signal, $X_a(\Omega)$, is illustrated in Fig. 4.2 (top).

2. The condition $\Omega_0 < \pi/T$ is true. This prevents aliasing. When this condition is violated, the output of the D/A will differ from the input to the A/D. An example is shown in Fig. 4.6. When $\Omega_0 < \pi/T$, the reconstruction is perfect, as shown in Fig. 4.6*b*. However, when $\Omega_0 > \pi/T$, $Y_a(\Omega)$ is corrupted by aliasing, as illustrated in Fig. 4.6*c*.

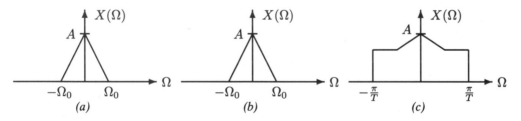

Figure 4.6. The effect of the sampling period on the performance of cascaded ideal A/D and D/A converters as shown in Fig. 5. (a) Bandlimited spectrum of the input to the A/D. (b) Output spectrum when $\Omega_0 < \frac{\pi}{T}$. (c) Output spectrum corresponding to a value of $\Omega_0 > \frac{\pi}{T}$.

The minimum sampling frequency for exact reconstruction, $f_s = \Omega_0/\pi$ samples/second, is called the *Nyquist rate*. If a signal is sampled at less than the Nyquist rate, the Fourier spectra in equation (4.2) overlap, as illustrated in Fig. 4.2 (bottom). This spectral overlap, which is called *aliasing*, prevents the signal from being recovered exactly. The reconstruction condition for a bandlimited signal $x_a(t)$ with a highest frequency component Ω_0 can be formalized as a sampling theorem and stated in several equivalent ways:

1. $x_a(t)$ may be reconstructed if $\Omega_0 < \pi/T$, or;
2. $x_a(t)$ must be sampled above its Nyquist rate, i.e., $f_s > \Omega_0/\pi$ samples/sec, or;
3. $x_a(t)$ must be sampled at a rate greater than twice the Nyquist frequency, i.e., $f_s > 2\Omega_0/(2\pi)$.

The exercises that follow provide illustrations of sampling, aliasing, and related issues associated with the discrete-time processing of continuous-time signals. Because of the difficulty of representing an analog waveform in a computer, a digital representation is used to simulate analog signals. In these exercises, several special functions will be used. These are rather restrictive and are only intended for use in the exercises in this chapter. The inputs and outputs, when appropriate, are displayed as continuous functions to give the appearance of an analog signal. A brief description of these special functions follows.

x aconvolve

The **x aconvolve** function forms the linear convolution of two simulated analog signals,

$$y_a(t) = x_a(t) * h_a(t).$$

It is executed by typing

$$\overbrace{\qquad}^{x_a(t)} \qquad \overbrace{\qquad}^{h_a(t)} \qquad \overbrace{\qquad}^{y_a(t)}$$

x aconvolve   (input1)   (input2)   (output)

x afilter

This special function is used to simulate analog filtering of the form

$$y_a(t) = h_a(t) * x_a(t)$$

where $x_a(t)$ and $y_a(t)$ are simulated analog input and output signals and $h_a(t)$ is the impulse response of the simulated analog filter that is provided for you in the file **hh**. To invoke the function, type

$$\overbrace{\qquad}^{x_a(t)} \qquad \overbrace{\quad}^{h_a(t)} \qquad \overbrace{\qquad}^{y_a(t)}$$

x afilter   (input)   hh   (output) M

M is the number of output samples. Since simulated analog signals are actually represented by discrete samples, this parameter is a necessary input. In these exercises 500 is a good value to specify for this number.

x alook

This function allows you to display simulated analog signals in the time domain. The signal is displayed as a continuous-time waveform where the time axis is measured in milliseconds (msec). It is invoked by typing

$$\overbrace{\qquad}^{x_a(t)}$$

x alook   (inputfile)

The function always displays the simulated analog signal in a 100-millisecond time interval. Pressing the "*g*" key allows you to turn the grid of tic marks on or off to suit your preference. The "*q*" key will terminate the display immediately and return you to DOS.

x ft

This function simulates the Fourier transform of an analog signal. It is invoked by typing

$$\overbrace{x_a(t)}$$

x ft $\overbrace{\text{(inputfile)}}^{x_a(t)}$

A display option menu will appear that will ask for display of the magnitude, log magnitude, phase, real part, imaginary part, magnitude and phase, or real and imaginary parts. The frequency axis should be interpreted as if it extends infinitely in both directions. Once the plot is displayed on the screen, three keys are available for use. Pressing the "*g*" key allows you to turn the grid of tic marks on or off to suit your preference. The "*esc*" key can be used to return you to the display option menu. The "*q*" key will terminate the display immediately and return you to DOS.

x sample

This function allows a simulated analog signal, $x_a(t)$, to be sampled. It simulates the sampling operation

$$x[n] = x_a(nT)$$

where T is the sampling period in milliseconds. The sampling period can be included on the command line or the program will prompt you for it. The program is invoked by typing

x sample $\overbrace{\text{(input)}}^{x_a(t)}$ $\overbrace{\text{(output)}}^{x[n]}$ T

The output sequence, $x[n]$, is truncated to 101 samples in all cases. This is done so that various sampled signals investigated in the exercises will all have the same number of samples and thus permit fair comparisons to be made. Unlike the other DSP functions, **x sample** is not equipped to handle input and output files with the same name.

x sti

This program simulates the "sample-to-impulse" conversion operation that transforms a sequence such as $y[n]$ into the weighted analog impulse train $y_c(t)$. It is invoked by typing

$$\overset{y[n]}{\overbrace{\text{(input)}}} \quad \overset{y_c(t)}{\overbrace{\text{(output)}}} \quad T$$

x sti

T is the integer number of milliseconds between impulses. If this parameter is omitted, the function will ask you for its value.

EXERCISE 4.1.1. **Ideal Sampling (Frequency Domain)**

This exercise considers the effects of sampling a continuous-time signal on the Fourier transform of the signal. The signal considered is the sine wave

$$x_a(t) = \sin \frac{2000\pi}{12} t.$$

To simulate this signal, generate 4000 samples of

$$\sin(0.2618n)$$

starting at $n = 0$, using the *sine wave* option in **x siggen**. Store the resulting sequence in a file.

(a) Use the function **x sample** to sample $x_a(t)$ with a sampling period, $T = 1$ msec. Call the resulting signal $x[n]$. Calculate and display the magnitude of the Fourier transform of $x[n]$ using **x dtft**. Carefully examine equation (4.2) and the plot of $|X(e^{j\omega})|$ to determine the frequency (in radians) of the discrete sinusoid. Recall that the DTFT of an ideal sinusoid is a pair of impulses. The location of these impulses on the frequency axis represents the frequency of the sinusoid.

(b) Repeat the sampling procedure described in part (a) for sampling periods of 4, 10, 15, 20, 24, 30, and 39 milliseconds. Display and sketch the plots of the magnitude response and record the frequencies of the discrete sinusoids in each case. How does the behavior of the peak locations of the DTFT magnitude change as the sampling period is increased? This behavior is due to *aliasing*.

(c) Based on the definition, determine the Nyquist rate for $x_a(t)$.

EXERCISE 4.1.2. **Ideal Sampling (Time Domain)**

This exercise looks at aliasing in the time domain. To generate $x_a(t)$ create 5200 samples of

$$\sin(0.1309n)$$

starting at $n = 0$ using the *sine wave* option in **x siggen**. Sample this waveform using **x sample** with a sampling period $T = 1$ millisecond. Use the function **x view** to display this sampled waveform.

(a) Repeat the sampling process described above for sampling periods of 2 and 4 milliseconds. Display and sketch the sampled waveforms. Determine the period and digital frequency (in radians) of each of the sampled signals. As the sampling rate is lowered, does the (digital) frequency of the signal increase or decrease? Remember that the sampling rate is the reciprocal of the sampling period.

(b) Now sample $x_a(t)$ with sampling periods of 50 and 51 milliseconds. Again use **x view2** to display the outputs and provide sketches of these signals. These are undersampled signals that contain aliasing. By comparing the periods to those generated in part (a), what does aliasing look like in the time domain?

EXERCISE 4.1.3. **Ideal Reconstruction (Frequency Domain)**

As we saw earlier, the process of ideal D/A conversion is equivalent to a two-step process. First, the sequence $y[n]$ is converted into a weighted continuous-time impulse train, $y_c(t)$. Then the impulse train is lowpass filtered to form $y_a(t)$. In this exercise D/A operations are examined in the frequency domain.

(a) Use the triangle wave option of **x siggen** to generate five periods of a triangle wave $y[n]$ with a period of 20 beginning at $n = 0$. Display and sketch the magnitude of the DTFT of $y[n]$.

(b) The first component of the D/A implements the "sample-to-impulse train" operation. This is simulated by the function **x sti** that converts the samples into impulses spaced T milliseconds apart. Use **x sti** with sampling period $T = 5$ to produce $y_c(t)$. Then using **x ft**, display and sketch $|Y_c(\Omega)|$. Note that $|Y_c(\Omega)|$ is periodic in Ω, but that **x ft** only evaluates it over a limited frequency range. The replication phenomenon that you observe is called *imaging*.

(c) The spectral replicas due to the sample-to-impulse conversion need to be removed. Use **x afilter** to implement the lowpass filter **hh** with the input signal $y_c(t)$ to produce $y_a(t)$. (Notice that you must specify $y_c(t)$ as the first argument of **x afilter** and **hh** as the second.) Specify 500 for the number of output points. Use **x ft** to display and sketch the magnitude of the frequency response of the lowpass filter and also $|Y_a(\Omega)|$. Recall that the sampling rate is $f_s = 1/T$ and label the frequency axis correctly in your sketch.

EXERCISE 4.1.4. **Commercial D/A Converters**

Commercially available D/A converters do not produce smooth continuous-time output signals, but instead create a piecewise-constant (staircase) output. This can be modeled by setting $h_a(t)$ in equation (4.3) equal to

$$h_a(t) = \begin{cases} 1, & 0 \le t \le T \\ 0, & \text{otherwise.} \end{cases}$$

The behavior of these D/As will be examined by simulation. Let the input to the D/A, $x[n]$, be 101 points of the sequence

$$\sin(\pi n/8 + \pi/10)$$

starting at $n = 0$. Use the *sine wave* option of **x siggen** to create $x[n]$.

(a) Let $x_s(t)$ be the output of a *sample-to-impulse* converter with $x[n]$ as the input. Use the **x sti** function with $T = 5$ to generate the simulated continuous-time signal $x_s(t)$. Display and sketch this signal using **x alook.** Let the output of the D/A be

$$\hat{x}_a(t) = x_s(t) * h_a(t).$$

To simulate $h_a(t)$, create a sequence of 5 ones using the *block* option in **x siggen** with $n = 0$ as the starting point. Display $h_a(t)$ using **x alook**. Then use the **x aconvolve** function to evaluate the convolution. Display and sketch $\hat{x}_a(t)$ using **x alook**.

(b) Now look at $|\hat{X}_a(\Omega)|$ using **x ft**. How does the stair-step output affect the spectrum of the signal?

(c) A simple lowpass filter can be used to filter $\hat{x}_a(t)$ so that it more closely resembles

$$\sin\left(\frac{\pi}{16}t + \frac{\pi}{10}\right).$$

Such a filter $h(t)$ is stored in the file **hh.** Use the **x ft** function to display the frequency response of this filter. Now use **x afilter** to lowpass filter $\hat{x}_a(t)$. Be sure to specify the input as the first argument and **hh** as the second argument on the command line. Specify 500 for the number of output points. Use **x alook** to display and sketch the result, $\hat{x}_a(t)$. Use **x ft** and **x alook** to display and sketch $|X_a(\Omega)|$.

Comment. In many real-world applications, the outputs of commercial D/As are passed through lowpass filters, as in this example, to smooth out the staircase-like structure and partially compensate for the D/A response.

EXERCISE 4.1.5. **DTFT of a Sampled Signal**

Do this exercise without the aid of your computer. Consider the case of a band-limited continuous-time signal with Fourier transform $X_a(\Omega)$ that has been sampled above its Nyquist rate so that no aliasing is present. Then

$$X(e^{j\omega}) = \frac{1}{T}X_a(\omega/T), \qquad -\pi \le \omega \le \pi.$$

Recall that $\Omega = \omega/T$.

(a) Consider the case where

$$X_a(\Omega) = \begin{cases} 1, & -2\pi(5000) \le \Omega \le 2\pi(5000) \\ 0, & \text{otherwise.} \end{cases}$$

Sketch $X(e^{j\omega})$ for the following sampling rates:

(i) $f_s = 15,000$ samples/sec.
(ii) $f_s = 30,000$ samples/sec.

    (iii) $f_s = 10,000$ samples/sec.

(b) When the signal is undersampled, the spectrum is aliased. In such cases, the general expression

$$X(e^{j\omega}) = \frac{1}{T} \sum_{r=-\infty}^{\infty} X_a \left(\Omega + \frac{2\pi r}{T} \right) \qquad \text{for } \Omega = \frac{\omega}{T}$$

relates the discrete-time and continuous-time Fourier transforms. Sketch $X(e^{j\omega})$ when $x_a(t)$ is sampled at the following rates:

    (i) $f_s = 8000$ samples/sec.

    (ii) $f_s = 5000$ samples/sec.

    (iii) $f_s = 3000$ samples/sec.

EXERCISE 4.1.6. **Sampling Non-Bandlimited Signals**

Many continuous-time signals encountered in the real world have spectra that are not strictly bandlimited, but are bandlimited only in an approximate sense. For example, a speech signal might contain some spectral energy above 10 kHz. But, if this high-frequency energy is small, the speech will not suffer noticeable distortion if it is removed by lowpass filtering. Thus, a good strategy to apply to signals of this type that need to be sampled is to determine a frequency above which the signal has negligible energy, and use this frequency to determine the sampling rate. When the sampling frequency is sufficiently high, the aliasing introduced will be small. This exercise addresses this aspect of sampling and does not involve the use of the computer.

Consider the analog input signal with Fourier transform

$$X_a(\Omega) = e^{-.01|\Omega|}.$$

Notice that $X_a(\Omega)$ decays exponentially as $|\Omega|$ increases. When this signal is sampled with any sampling rate, f_s, aliasing will be introduced—the larger the value of f_s, the smaller the amount of aliasing.

(a) If the spectral distortion due to aliasing is not permitted to exceed 1% at any frequency, what is the minimum sampling rate that is permissible?

(b) Now assume that an analog lowpass filter (often called an antialiasing filter) is used to prefilter the signal. This is a common practice because it removes the high-frequency spectral components so that they cannot alias onto the lower frequencies. If the maximum allowable spectral distortion is still 1% at any frequency, what is the minimum sampling rate that is permissible?

4.2 QUANTIZATION

Processors for discrete-time signals must be capable of adding, multiplying, and storing sample values. This means that each sample value must be representable in a form that requires only a finite number of bits. There are a variety of ways for doing this, but all of them share the property that only a limited number of sample amplitudes can be represented. Once a sample value has been measured, it must be approximated by one of these representable values. This process is called *quantization*.

Two examples of quantizer characteristics are shown in Fig. 4.7. The input to the quantizer is $x[n]$ and the output is $\hat{x}[n]$. These are both examples of *uniform* quantizers because the differences between successive quantization levels are equal to a constant value, Δ, which is called the *step size*. The difference between the unquantized and the quantized sample values

$$e[n] = \hat{x}[n] - x[n] \qquad (4.9)$$

is called the *quantization error.* If $X_{\min} < x[n] \leq X_{\max}$, then

$$|e[n]| \leq \frac{\Delta}{2}.$$

When $x[n]$ extends beyond these limits, the error can be much larger. This condition is called *saturation* or *overload*. Saturation can be minimized by carefully controlling the amplitude of the input to the A/D converter.

In general, the larger the number of levels, the more accurate the representation or equivalently the smaller the quantization error. Minimizing quantization error is an important issue in signal processing. A common measure of quantization error (noise) is the *signal-to-noise ratio (SNR)*. Rearranging equation (4.9) gives

$$\hat{x}[n] = x[n] + e[n].$$

The quantized signal $\hat{x}[n]$ is equal to the true signal plus the quantization noise. The signal-to-noise ratio (SNR) is defined as

$$\text{SNR} = 10 \log_{10} \frac{\sum_n x^2[n]}{\sum_n e^2[n]} \qquad \text{(in dB)}$$

where *dB* denotes *decibel*. The quantizer at the top of Fig. 4.7 is called a *midriser uniform quantizer*. It does not have a level corresponding to the value zero. The lower quantizer is called a *midtread quantizer*. When $X_{\min} = -X_{\max}$ a quantizer with an even number of levels will be a midriser and one with an odd number of levels will be a midtread. In practice, midtread quantizers are popular because they include the zero-value quantization level. The quantizers associated with most A/D converters have 2^B levels, where B is the number of bits that their output registers can hold. Since 2^B is even, midtread quantizers cannot be symmetrical about zero. The common convention is for there to be one extra negative quantization level. In the exercises that follow, quantization is examined in further detail.

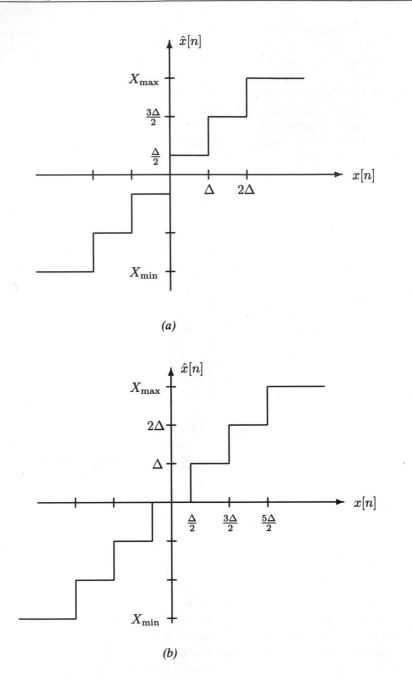

Figure 4.7. Two examples of uniform quantizers. (a) Six-level midriser quantizer. (b) Seven-level midtread quantizer.

***EXERCISE 4.2.1.* Quantization**

To illustrate quantization generate the ramp signal $r[n] = n/999$ in the range -1000 to $+1000$. Do this by first using **x siggen** to create a ramp with slope $1/999$, length 1000, and starting point zero. Next use **x reverse** to rotate the ramp about $n = 0$. Multiply the rotated ramp by -1 using **x gain** and add it to the original ramp using **x add**. Use **x look** to display $r[n]$ and verify that it is a straight line in the proper range.

If $r[n]$ is quantized to 3 bits/sample, then 8 unique amplitude levels can be represented. The first of these might be assigned the binary code 000, the second 001, and so on with the last being assigned the code 111.

(a) Use the function **x quantize** to quantize $r[n]$ to eight levels. Set the minimum and maximum quantizer levels (X_{min} and X_{max}) to -1 and $+1$, respectively. The quantized signal will be called $\hat{r}[n]$. Display $r[n]$ and $\hat{r}[n]$ one above the other using **x look2**. Now use **x slook2** to display these plots superimposed on each other and provide a sketch.

Examine the quantizer relationship shown in Fig. 4.7. Has $r[n]$ been quantized by a midriser or by a midtread quantizer? What is the step size, Δ?

(b) Examine the distortion that was introduced in part (a) as a result of quantization and sketch the result. Do this by explicitly computing the error signal,

$$e[n] = \hat{r}[n] - r[n]$$

using **x subtract** and displaying $e[n]$ using **x look**. What is the maximum value of $e[n]$?

(c) The quantization error in our example can be reduced by setting $X_{min} = -.875$ and $X_{max} = .875$. Requantize $r[n]$ as before using these limits; display and sketch the results. Compute and display the error signal, $e[n]$, and determine the maximum error.

This illustrates that the best performance is *not* achieved when the minimum and maximum amplitudes of the signal and quantizer are the same. If the minimum and maximum amplitudes of a signal $x[n]$ are $-|A|$ and $+|A|$, respectively, and L is the number of desired quantization levels, what should X_{min} and X_{max} be to achieve the lowest quantization error? Express your answer in terms of L.

(d) Quantize $r[n]$ to nine levels instead of eight. Determine and record the values of X_{min} and X_{max} that will yield the minimum quantization error. What is the value of the maximum quantization error? Display and sketch $r[n]$ and $\hat{r}[n]$ as before.

Next create a 21-point sequence $x[n] = 5\sin \pi/10$ in the range $0 \leq n < 21$ using **x siggen**. Quantize $x[n]$ to eight levels using **x quantize**. Again select X_{min} and X_{max} to minimize quantization error. Use **x view2** to simultaneously display $x[n]$ and the quantized signal; sketch these plots. What is the value of Δ?

Real A/D converters must quantize the sampled analog input so that it can be represented on a computer. In practice, 12- to 16-bit quantization provides a more than adequate signal representation for most applications. As shown in the previous exercise, the effects of quantizing with as few as 3 bits (or 8 levels) still preserves most of the waveform's time structure.

EXERCISE 4.2.2. **Quantization Effects**

The number of quantization levels in an A/D converter varies somewhat depending on the application. This exercise will investigate the effects of varying the number of quantization levels.

A system composed of cascading A/D and D/A converters can be simulated by the following macro, which will be called **sys.bat**:

```
echo off
x sample    %1    dummy    1
x quantize    dummy    dummy1    %3    %4    %5
x sti    dummy1    dummy2    1
x afilter    dummy2    hh    dummy    150
x extract    dummy    %2    100    10    0
del dummy
del dummy1
del dummy2
echo on
```

Using your text editor, copy this macro onto your computer. The macro has five command-line arguments: the input, the output, the minimum quantizer limit, the maximum quantizer limit, and the number of quantizer levels. The macro receives, as its input, a simulated analog signal file, $x_a(t)$, and returns another simulated analog signal file, $y_a(t)$. The macro is executed by typing:

$$\overbrace{x_a(t)}\qquad\overbrace{y_a(t)}$$

sys   (inputfile)   (outputfile)   min   max   levels

Since no processing is performed, the input and output should be the same except for the distortion introduced by aliasing, quantization, and the non-ideal phase response of the filter. The purpose of the **x extract** statement is to remove a time shift that is introduced by the filter. The system assumes that time is measured in milliseconds and that the sampling period is 1 msec. The output signal length is assumed to be 150.

In this exercise, you will work with a simulated linearly increasing analog triangular wave, $x_a(t)$. To create this signal, use **x siggen** to generate five periods of a triangle wave with a period of 20 samples, amplitude 0.01, and starting point $n = 0$. Next multiply this signal by a 100-point ramp function that can be created using **x siggen**. This produces a triangular wave, $x_a(t)$, with a linearly increasing envelope.

(a) Using **x alook**, display and sketch $x_a(t)$ and verify that it is correct. What are the appropriate values of min and max for the **sys.bat** macro given the amplitude range of $x_a(t)$?

(b) In this part, the number of levels will be varied in each experiment. Using **x alook,** display and sketch the first 100 points of the **sys** output when the number of levels is 1024. How many bits are required for this representation? Sketch the output when 16 levels, 4 levels, and 2 levels are used. Notice that the greatest distortion is due to aliasing and filtering errors from sampling. However, distortions due to quantization are also evident. For this particular signal, what seems to be the smallest number of bits needed to yield acceptable performance?

EXERCISE 4.2.3. **Pulse Code Modulation (PCM)**

PCM is the simplest method for discrete-time signal representation. It quantizes each of the sampled values to a finite number of levels and then represents each of these levels with a unique binary codeword. PCM serves as the basis for digitally encoding waveforms for transmission or storage. If the waveform is sampled $1/T$ times per second and if the samples can be adequately represented using a B-bit quantizer, then the waveform can be transmitted digitally over a communication channel that has a capacity of B/T bits/sec.

The quantization inherent in PCM degrades the quality of the signal and thus lowers its SNR. The impact of different numbers of quantization levels on the SNR is investigated in this exercise.

(a) Let $x[n]$ be the signal contained in file **sig1**. Examine the signal using **x look** to determine maximum and minimum values that will avoid any saturation, then use **x quantize** to quantize $x[n]$ to 4 bits (16 levels) to form $\hat{x}[n]$. Compute the SNR using the **x snr** function and record the result. Note that $x[n]$ and $\hat{x}[n]$ are the arguments of the function. Next quantize $x[n]$ to 6, 8, and 10 bits and record the SNRs. Determine a functional (equation) relationship between the SNR and the number of bits used for the quantization.

(b) Using your result from (a), what value would you predict for the SNR if a 16-bit quantizer were used?

EXERCISE 4.2.4. **Differential PCM (DPCM)**

There are a number of alternatives to PCM (which was introduced in the previous exercise) that may permit less quantization error at the same bit rate, or, alternatively may allow the bit rate to be reduced without degrading the SNR. One of these alternatives is illustrated in this exercise. It has some of the flavor of a waveform coding system known as differential PCM (DPCM). The simple differential system considered here consists of three subsystems as shown in Fig. 4.8: an FIR filter (on the left); a uniform quantizer, shown in the center; and an inverse filter that compensates for the response of $h[n]$, shown on the right. The linear and time-invariant FIR filter is defined by the difference equation

$$y[n] = x[n] - ax[n-1].$$

Quantization is performed on $y[n]$ resulting in $\hat{y}[n]$. This signal may be thought of as the compressed signal or encoded signal. The goal is to represent this signal with as few bits as possible.

Figure 4.8.   A simple differential PCM system.

The purpose of the third subsystem in the DPCM coder expands the coded signal $\hat{y}[n]$ into a signal $\hat{x}[n]$ that approximates the original input $x[n]$ as closely as possible. This system is the first-order IIR filter that is the inverse of the FIR filter. It is defined by the difference equation

$$\hat{x}[n] = a\hat{x}[n-1] + \hat{y}[n].$$

As you will see, the overall system can encode $x[n]$ more accurately than a PCM system.

Note that since both digital filters can be implemented as difference equations, the **x lccde** function can be used to evaluate both $y[n]$ and $\hat{x}[n]$. Let the length for the **x lccde** output file be 128 samples.

(a) Calculate the sequence $y[n]$ when $x[n]$ is the signal in **sig2** for $a = .8$. Quantize it to 4-bits using **x quantize**. You will need to examine the signal $y[n]$ with **x look** to set the minimum and maximum levels properly to avoid saturation. Reconstruct the signal $\hat{x}[n]$ and evaluate the signal-to-noise ratio (SNR) of the overall system using **x snr**. For comparison determine the SNR of a 16-level PCM (see previous exercise) encoder applied to $x[n]$.

(b) Determine how many quantization levels would be required in a PCM encoder that has the same SNR as the DPCM coder designed in part (a).

(c) Write a macro to implement your coder. Let the parameter a be variable and measure the SNR for different values of a using the signal **sig2**. For what a is the SNR maximized?

The structure of actual DPCM systems is somewhat different than that described here, and the filters ($h[n]$) are often optimized to the statistics of the signals being encoded.

4.3 SAMPLING-RATE CONVERSION

The higher the sampling rate the more samples must be processed. Consequently, for processing efficiency the sampling rate should be reduced whenever possible. Such a situation arises when preliminary digital processing, involving operations such as low-pass filtering, reduces a signal's bandwidth, producing an output signal that is over-sampled. One way to reduce the sampling rate is to convert the digital signal back into analog form using a D/A converter and then resample it, but this is expensive. It requires additional components and the A/D and D/A converters are often the least accurate components in a complete realization. A simpler approach is to do the sampling-rate conversion in the discrete-time domain.

The sampling rates of a discrete-time signal are changed using operators called *decimators* and *interpolators*. A system for decimating the sampling rate by a factor of N is shown in Fig. 4.9. A decimator consists of a cascade of a lowpass filter, $H(e^{j\omega})$, and a downsampler (or subsampler), denoted by $\downarrow N$ in Fig. 4.9. The downsampler operates on successive blocks of N samples by retaining the first sample in each block and discarding the rest, i.e.,

$$y[n] = x[Nn].$$

This reduces the number of samples in the sequence by a factor of N. The usage of the term *decimator* is not consistent in the signal processing literature. Very often it is used to refer to the downsampling operation. However, we will consistently use *decimation* to refer to the combination of lowpass filtering and downsampling as shown in Fig. 4.9.

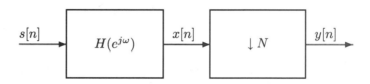

Figure 4.9.   A decimator for reducing the sampling rate by a factor of N.

In the frequency domain, the input and the output of the downsampler are related by

$$Y(e^{j\omega}) = \frac{1}{N} \sum_{r=0}^{N-1} X(e^{j(\omega/N + (2\pi r)/N)}). \tag{4.10}$$

Downsampling introduces aliasing into the DTFT of $x[n]$. It is therefore important to bandlimit the input of the downsampler to the frequency range $|\omega| < \pi/N$. This is the role of the lowpass filter $H(e^{j\omega})$ that precedes the downsampler. It has unity gain and a cutoff frequency of π/N.

The dual operation is called *interpolation*. It is the process by which a signal sampled at one sampling rate is converted to one with a rate that is L times higher. Figure 4.10 shows an implementation of a rate L interpolator. It is also a two-stage operation. The first subsystem, which is called an *upsampler*, inserts $L - 1$ zeros between each sample in the input. The second subsystem is a lowpass filter with a gain of L and a cutoff frequency of π/L. As in the case of decimation, the term *interpolation* is sometimes used in the literature to refer to just the upsampling operation, but we will adhere to the definition of Fig. 4.10 in this text.

The frequency-domain relationships that describe interpolation are

$$R(e^{j\omega}) = X(e^{jN\omega})$$
$$Y(e^{j\omega}) = R(e^{j\omega})H(e^{j\omega}) = X(e^{jN\omega})H(e^{j\omega}).$$

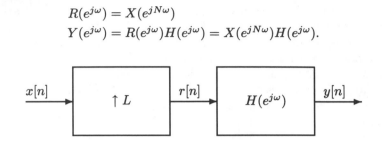

Figure 4.10.   A system for interpolating by a factor of L.

EXERCISE 4.3.1. **Downsampling**

The downsampler is the basic component for sampling rate reduction. To examine its operation more closely, consider the input signal $x[n]$ where

$$x[n] = \cos\left(\frac{\pi}{32}n\right).$$

Let this signal be the input to a downsampler with a decimation factor of $M = 3$ and let the output be $y[n]$.

(a) Based on the definition of decimation, determine an explicit expression for $y[n]$.

(b) The DTFT of a cosine is a pair of impulses in the frequency domain. Using equation (4.10) sketch $|Y(e^{j\omega})|$. This does not involve the use of a computer.

(c) Now consider a time-limited input sequence $x_T[n]$ consisting of 512 samples of $x[n]$ taken over the range $0 \le n < 512$. Create $x_T[n]$ using **x siggen**. Using the **x dnsample** function compute the output $y_T[n]$ of the decimator. Use the **x truncate** function to limit both $x_T[n]$ and $y_T[n]$ to 128 samples so that you are comparing signals of the same length. Use **x dtft** to display and sketch the discrete-time Fourier transform magnitudes of these signals and compare them with the results found in part (b). Note that for finite length segments of sinusoids the spectrum is not truly impulsive, but is approximately so.

(d) Is the downsampling operation linear? Justify your answer analytically.

(e) Is the downsampling operation shift invariant? Justify your answer analytically.

EXERCISE 4.3.2. **Aliasing in Downsampling**

This problem examines the effects of downsampling in the frequency domain. Use **x siggen** to create 2450 samples of $x[n]$ beginning at $n = 0$ where

$$x[n] = \cos\frac{\pi n}{11}.$$

(a) Downsample $x[n]$ by the following factors and truncate the results to 100 samples using **x dnsample** and **x truncate**.

(i) $N = 4$

(ii) $N = 18$

(iii) $N = 20$

(iv) $N = 24$.

For each case display and sketch the transform magnitude using **x dtft**. In which downsampled signals is aliasing present? Assume that $x[n]$ was obtained by taking samples of the output of an ideal A/D with a sampling rate $f_s = 100$ Hz. What sampling rates could have been used with the analog signal in order to produce each of the downsampled signals?

EXERCISE 4.3.3. **Upsampling**

An upsampler inserts zeros between the sample values of a sequence. To explore the properties of this operation, use **x siggen** to generate 66 points of the sequence $x[n]$ in the range $0 \leq n \leq 65$ where

$$x[n] = \cos \frac{\pi}{5} n.$$

(a) Using **x upsample**, examine the sequences obtained by

(i)   upsampling $x[n]$ by a factor of 2,

(ii)   upsampling $x[n]$ by a factor of 10.

Sketch these sequences in the time domain.

(b) Truncate each of the sequences that you generated in (a) to 128 points using **x truncate**. Display and sketch these sequences in the time domain and the frequency domain using **x view** and **x dtft**. This phenomenon that you observe whereby spectral copies are replicated is called *imaging*.

(c) Is the upsampling operation linear? Explain.

(d) Is it shift invariant? Explain.

EXERCISE 4.3.4. **Commutativity of Downsampling and Upsampling**

Systems are said to be *commutative* if the result of applying multiple operations to a signal is independent of the order in which they are applied. In general the downsampling and upsampling operators are not commutative. Downsampling by a factor of N followed by upsampling by a factor of M is not equivalent to upsampling by M followed by downsampling by N. For certain values of M and N, however, downsampling and upsampling are commutative.

(a) Determine a pair of factors, M and N, such that downsampling and upsampling are commutative or order independent. Exclude the trivial case for which $N = M = 1$.

(b) What condition must N and M satisfy in order for the upsampling and downsampling operations to be commutative? Use the computer to test examples that support your conclusion.

EXERCISE 4.3.5. **Decimation and Aliasing**

You have seen that aliasing can occur as a result of downsampling. If a signal is not bandlimited to the range $-\pi/N < \omega < \pi/N$, then downsampling by N will cause undesired aliasing. To illustrate this point consider the signal $gg[n]$ that is stored in the file **gg**.

(a) Display and sketch the DTFT of $gg[n]$. Now downsample the signal by a factor of 3 using **x dnsample** and sketch the DTFT of the result using **x dtft**. Verify this result graphically by adding together the appropriate spectral replicas of $GG(e^{j\omega})$ and scaling the result.

(b) Now decimate the signal by a factor of 3 by first filtering it with a 201-point FIR lowpass filter with a cutoff frequency $\pi/3$ that you design using the **x fdesign** function with the Hamming window option. Convolve the impulse response of the filter with the signal $gg[n]$ using **x convolve** and downsample the result using **dnsample**. Display and sketch the DTFT magnitude using **dtft**.

EXERCISE 4.3.6. **Interpolation**

Interpolation is achieved by upsampling and filtering with a lowpass filter. To illustrate this, use **x siggen** to generate the 64-point chirp signal with *amplitude* = 1, $\alpha_1 = 0.4$, $\alpha_2 = 0.01$, *phi* = 0, and starting point $n = 0$.

(a) Display and sketch the DTFT magnitude of the chirp signal.

(b) Upsample and interpolate the chirp signal by a factor of 3. The functions **x upsample**, **x convolve**, **x gain**, **x extract**, and **x fdesign** may be used for implementation and design of the system. Try using a 64-point lowpass filter designed with the *Hamming window* option and cutoff frequency $\pi/3$. Look at the signal in the frequency domain and describe its relationship to the original.

EXERCISE 4.3.7. **Fractional Sampling Rate Conversion**

Sometimes sampling-rate conversions are needed between sampling rates that are not integer multiples of one another. One solution to this problem is to first interpolate by a factor of M and then decimate by a factor of N. This permits a fractional sampling-rate change by a factor of M/N to be made. A system for doing this is shown in Fig. 4.11. The lowpass filters from the interpolator and the decimator can be replaced by a single filter with a gain of M and a cutoff frequency of $\pi/\max(M, N)$, i.e., the cutoff is the smaller of π/M or π/N.

(a) Create a macro for implementing a sample-rate conversion by a general factor of M/N. The function **x fdesign** with the Hamming window option can be used to design the lowpass filter. Use a filter length of 31. In addition use **x lshift** to shift the impulse response by 15 samples so that it will have zero phase.

(b) Verify the operation of your system by generating the signal

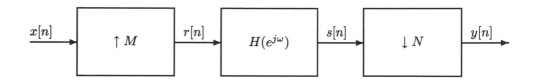

Figure 4.11.   A cascade of an interpolator and a decimator to achieve a sample-rate conversion of M/N.

$$x[n] = \cos\frac{\pi}{30}n + \cos\frac{\pi}{5}n$$

over the range $0 \le n < 128$ using **x siggen**. Use your macro to change the sampling rate by a factor of 3/4, and denote the resulting sequence by $y[n]$. Use **x look** to display $x[n]$ and $y[n]$ as continuous waveforms. Sketch these plots. Also examine $x[n]$ and $y[n]$ using **slook2**, which will superimpose the waveforms on the same plot. Observe the change in frequency due to the rate change.

(c) Repeat part (b) for the case where $M = 6$ and $N = 8$. Use a 31-point lowpass filter designed as before. Note that the ratio is the same as in part (b), but the filter, upsampler, and downsampler are different. Use **x look2** to display the outputs from parts (b) and (c) one above the other and sketch these plots. Redesign the 31-point filter and perform the same rate change using $M = 24$, $N = 32$. Sketch these results as before. Explain why the results are different.

(d) Apply your macro to the signal, $y[n]$, that you produced in (b) with a rate change factor of 4/3. Compute and sketch the DTFT magnitude of the new signal and compare it to that of the original. Was the inverse rate change successful in restoring the signal to its original sampling rate?

EXERCISE 4.3.8. **Fractional Sample Delay**

The system shown in Fig. 4.12 implements a delay of $1/N$, which is less than one sample.

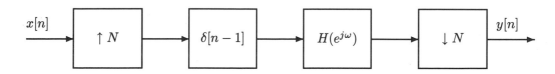

Figure 4.12.   A system for implementing a delay of $1/N$ samples.

(a) Create a macro for implementing a fractional sample delay of 1/2 sample. Use the function **fdesign** with the *Hamming window* option to design the 31-point lowpass filter.

(b) Apply your macro to 64 samples of the sinusoid

$$x[n] = \cos \frac{\pi}{16} n$$

in the range $0 \leq n < 64$. Sketch the resulting time signal.

(c) Generate a unit sample, $\delta[n]$, using **x siggen** and delay it by 1/2 sample. Sketch and explain your result.

(d) Repeat part (c) for a 64-point pulse using the *block* option in **x siggen**.

The z-Transform and Flow Graphs

5.1 INTRODUCTION TO THE z-TRANSFORM

The z-transform, an important tool for discrete-time signal and system analysis, can be viewed as the discrete-time analog of the Laplace transform. It provides another domain in which signals and systems can be examined. The z-transform is very closely related to the discrete-time Fourier transform and finds its principal use in the analysis and manipulation of linear, constant coefficient difference equations.

The *two-sided* or *bilateral* z-transform is defined as the summation

$$X(z) = \sum_{n=-\infty}^{\infty} x[n]z^{-n}. \tag{5.1}$$

Although other definitions are useful for some purposes, this is the most useful in digital signal processing and, consequently, is the only one that we will use in this text. The summation in equation (5.1) is a power series in the complex variable z^{-1}, the coefficients of which are the samples of the sequence $x[n]$. As with any complex variable, values of z can be associated with points in a plane. A point z, plotted in the z-plane as in Fig. 5.1, has as its ordinate the imaginary part of z and as its abscissa the real part of z. It is also convenient to express points in the z-plane in polar coordinates in which a value is parameterized by its distance from the origin and its angle measured relative to the real axis.

Two important properties of the z-transform are apparent from its definition in equation (5.1). First the transform is linear. If

$$x[n] \iff X(z)$$

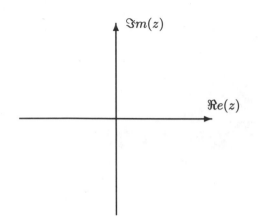

Figure 5.1.   The z-plane for representing values of the complex variable z.

and

$$v[n] \Longleftrightarrow V(z),$$

then

$$ax[n] + bv[n] \Longleftrightarrow aX(z) + bV(z).$$

The second property allows us to deal with shifts in $x[n]$. Delaying $x[n]$ by one sample multiplies its z-transform by z^{-1}, i.e.,

$$x[n-1] \Longleftrightarrow z^{-1}X(z). \tag{5.2}$$

Thus in the z-domain multiplication by z^{-1} corresponds to a delay operation. Similarly, delaying a sequence by m samples multiplies its z-transform by z^{-m}. The shift, m, can be either positive or negative.

The transform pair

$$\delta[n] \Longleftrightarrow 1$$

follows immediately from the definition of the z-transform. Evaluating the z-transform of an arbitrary finite length sequence becomes trivial when this relation is used with the linearity and shift properties. For example, the z-transform of

$$x[n] = 2\delta[n+1] + \delta[n] + 2\delta[n-1] + 3\delta[n-2] + 8\delta[n-3]$$

is simply

$$X(z) = 2z + 1 + 2z^{-1} + 3z^{-2} + 8z^{-3}.$$

Sequences that are infinitely long are transformed the same way, but it is frequently possible to write these z-transforms in closed form. This is true, for example, for the impulse responses of linear, time-invariant systems that are described by linear, constant coefficient difference equations. As an example, consider the z-transform of the signal

$$h[n] = \alpha^n u[n].$$

$H(z)$ is found by substituting into the definition in equation (5.1) to obtain

$$H(z) = \sum_{n=-\infty}^{\infty} \alpha^n u[n] z^{-n} = \sum_{n=0}^{\infty} (\alpha z^{-1})^n. \tag{5.3}$$

The geometric series formulas that were discussed in Chapter 2, in particular,

$$\sum_{n=0}^{\infty} a^n = \frac{1}{1-a}, \tag{5.4}$$

can be used to express this summation in closed form provided that

$$|a| < 1. \tag{5.5}$$

If $|a|$ is not less than 1, the summation does not converge (i.e., its value is infinite).

Applying this formula to equation (5.3) in our example gives

$$H(z) = \frac{1}{1 - \alpha z^{-1}}, \qquad |z| > |\alpha|.$$

For this to be valid, the convergence condition of equation (5.5) must hold. This implies that this z-transform is defined only for those values of z for which $|\alpha z^{-1}| < 1$ or $|z| > |\alpha|$. These values fall in the portion of the z-plane that is shaded in Fig. 5.2. This region is called the *region of convergence* (ROC). The specification of the ROC is part of the specification of the z-transform. This is made clear in the next example.

As a second example consider the z-transform of

$$x[n] = -\beta^n u[-n - 1].$$

Inserting the sequence into the defining summation and manipulating it into the form of a geometric series produces the result

$$X(z) = -\sum_{n=-\infty}^{-1} \beta^n z^{-n} = -\sum_{n=\infty}^{1} \beta^{-n} z^n$$

$$= -\sum_{n=1}^{\infty} \beta^{-n} z^n = -\beta^{-1} z \sum_{n=0}^{\infty} (\beta^{-1} z)^n. \tag{5.6}$$

In closed form this becomes

$$X(z) = -\frac{\beta^{-1} z}{1 - \beta^{-1} z} = \frac{1}{1 - \beta z^{-1}}.$$

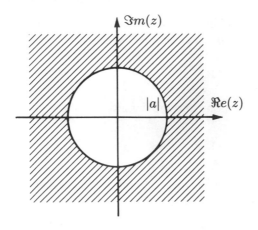

Figure 5.2.   The region of convergence in the z-plane for the first example.

The region of convergence is seen to be $|z| < |\beta|$.

These two examples illustrate an important fact. The z-transform of the left-sided sequence $-\beta^n u[-n - 1]$ is almost identical to the z-transform of the right-sided sequence $\beta^n u[n]$. Only their regions of convergence are different. For the left-sided sequence the ROC is $|z| < \beta$; for the right-sided sequence it is $|z| > \beta$.

From this discussion, three useful classes of signals can be identified:

1. **Right-sided sequences** of the form

$$x[n] = A_1\alpha_1^n u[n] + A_2\alpha_2^n u[n] + A_3\alpha_3^n u[n] + \cdots . \qquad (5.7)$$

They have z-transforms of the form

$$X(z) = \frac{A_1}{1 - \alpha_1 z^{-1}} + \frac{A_2}{1 - \alpha_2 z^{-1}} + \frac{A_3}{1 - \alpha_3 z^{-1}} + \cdots .$$

$A_1, A_2, A_3, \ldots$ are constants. The ROC consists of those values of z for which the z-transform of each of the exponential terms in equation (5.7) converges. This is the intersection of the individual ROCs for each term. Consequently, right-sided sequences will have regions of convergence that are the exteriors of circles in the z-plane, i.e., $|z| > |\alpha_m|$ where $|\alpha_m| = \max(|\alpha_k|, \quad k = 1, 2, \ldots)$ is the exponential of largest magnitude in equation (5.7).

2. **Left-sided sequences** of the form

$$x[n] = -B_1\beta_1^n u[-n - 1] - B_2\beta_2^n u[-n - 1] - B_3\beta_3^n u[-n - 1] - \cdots . \quad (5.8)$$

They have z-transforms of the form

$$X(z) = \frac{B_1}{1 - \beta_1 z^{-1}} + \frac{B_2}{1 - \beta_2 z^{-1}} + \frac{B_3}{1 - \beta_3 z^{-1}} + \cdots .$$

$B_1, B_2, B_3, \ldots$ are constants. Here the ROC is the interior of a circle, corresponding to the intersection of the regions of convergence of the individual terms. Convergence occurs for $|z| < |\beta_m|$ where $|\beta_m| = \min(|\beta_k|, \quad k = 1, 2, \ldots)$ is the exponential of smallest magnitude in equation (5.8).

3. **Finite duration sequences.** These sequences have both a beginning and an end.

$$x[n] = c_0 \delta[n - L] + c_1 \delta[n - L - 1] + \cdots$$
$$+ c_{J-1} \delta[n - L - (J - 1)] + c_J \delta[n - L - J] \qquad (5.9)$$

They have z-transforms of the form

$$X(z) = c_0 z^{-L} + c_1 z^{-L-1} + \cdots + c_{J-1} z^{-L-(J-1)} + c_J z^{-L-J}.$$

In this case the ROC is the entire z-plane, since a finite sum of finite quantities is finite. The only points in the z-plane that may be excluded from the ROC are $z = 0$ and $z = \infty$. At these values some of the individual terms can become infinite.

A more general class of sequences is the class of two-sided sequences that can be expressed as composites of right-sided, left-sided, and finite duration sequences. The most useful z-transforms encountered in digital signal processing can be expressed as rational functions of the form

$$X(z) = \frac{B(z)}{A(z)} = z^{-L} \frac{b_0 + b_1 z^{-1} + \cdots + b_M z^{-M}}{1 + a_1 z^{-1} + \cdots + a_N z^{-N}}$$

where $X(z)$ is a ratio of polynomials in z^{-1} and $N, M \geq 0$. The roots of the numerator polynomial (interpreted as a function of z, not z^{-1}) are called *zeros*, and the roots of the denominator polynomial are called *poles*. When a polynomial of the form

$$b_0 + b_1 z^{-1} + \cdots + b_M z^{-M}$$

is written as

$$(1 - \beta_1 z^{-1})(1 - \beta_2 z^{-1}) \cdots (1 - \beta_M z^{-1}),$$

its roots are seen to occur at $z = \beta_1, z = \beta_2, \ldots, z = \beta_M$. The first example discussed earlier had a z-transform of the form

$$H(z) = \frac{1}{1 - \alpha z^{-1}}, \qquad |z| > |\alpha|.$$

This has a zero at $z = 0$ and a pole at $z = \alpha$. The second example, which has a z-transform with the same functional form but a different region of convergence, also has a zero at $z = 0$ and a pole at $z = \beta$.

It is convenient to plot the poles and zeros in the z-plane. This is called a *pole/zero plot*. The zeros and poles are typically plotted using o's for the zeros and ×'s for the

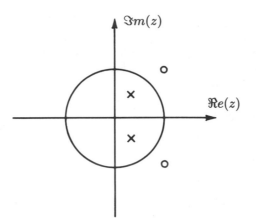

Figure 5.3.   An example of a pole/zero plot.

poles. Figure 5.3 shows an example of a pole/zero plot for a signal with two complex poles and two complex zeros.

The locations of the poles influence the possible choices for the regions of convergence in the following ways:

1. The ROC of the *z*-transform has poles on its boundaries. A pole may never be included in the ROC, since at any value of *z* where the denominator polynomial is zero, $X(z)$ is not finite. The ROC may be either the interior of a circle, an annular ring bounded by two concentric circles, or the exterior of a circle.

2. If a signal is right-sided, its *z*-transform converges on the exterior of the circle that is defined by the pole with the largest magnitude. If a signal is left-sided, its *z*-transform converges on the interior of the circle that is defined by the pole with the smallest magnitude.

The *z*-transform of the impulse response of a linear time invariant (LTI) system is called its *system function*. The system function (with its ROC) is a unique representation for the system. Since the impulse response of a causal system is right-sided, its system function will converge on the exterior of the circle defined by the largest pole. The ROC of the system function of a stable system must include the unit circle. Inclusion of the unit circle implies that $|z|$ can equal unity in the *z*-transform summation. This, in turn, implies that the impulse response is absolutely summable. Thus we can determine whether an LTI system is stable by inspecting its ROC.

The *z*-transform has many well-known properties that make it a valuable and powerful analysis tool. Table 5.1 presents an abbreviated list of these properties. In this table the constants α and β are complex and the constant n_0 is an integer. Each of these properties can be derived from the *z*-transform definition. The regions of convergence of the transformed signals are not shown. They are related to the ROC of $X(z)$, but in most cases are different from it.

An abbreviated list of common *z*-transform pairs is given in Table 5.2.

Table 5.1.   Properties of the z-Transform

Signal		z-Transform
$x[n]$	$\Longleftrightarrow$	$X(z)$
$x[n - n_0]$	$\Longleftrightarrow$	$z^{-n_0} X(z)$
$\alpha^n x[n]$	$\Longleftrightarrow$	$X(z/\alpha)$
$e^{-j\alpha n} x[n]$	$\Longleftrightarrow$	$X(e^{j\alpha} z)$
$x[-n]$	$\Longleftrightarrow$	$X(z^{-1})$
$nx[n]$	$\Longleftrightarrow$	$-z \dfrac{d}{dz} X(z)$
$x[n] * h[n]$	$\Longleftrightarrow$	$X(z)H(z)$

Table 5.2.   Common z-Transform Pairs

Signal		z-Transform
$x[n]$	$\Longleftrightarrow$	$X(z)$
$\delta[n]$	$\Longleftrightarrow$	$1 \quad 0 \le |z| \le \infty$
$\delta[n - n_0]$	$\Longleftrightarrow$	$z^{-n_0} \quad |z| \neq 0$ or $|z| \neq \infty$
$u[n]$	$\Longleftrightarrow$	$\dfrac{1}{1 - z^{-1}} \quad |z| > 1$
$\alpha^n u[n]$	$\Longleftrightarrow$	$\dfrac{1}{1 - \alpha z^{-1}} \quad |z| > |\alpha|$
$-\alpha^n u[-n - 1]$	$\Longleftrightarrow$	$\dfrac{1}{1 - \alpha z^{-1}} \quad |z| < |\alpha|$
$n\alpha^n u[n]$	$\Longleftrightarrow$	$\dfrac{\alpha z^{-1}}{(1 - \alpha z^{-1})^2} \quad |z| > |\alpha|$
$-n\alpha^n u[-n - 1]$	$\Longleftrightarrow$	$\dfrac{\alpha z^{-1}}{(1 - \alpha z^{-1})^2} \quad |z| < |\alpha|$
$\alpha^n \cos(\omega_0 n) \, u[n]$	$\Longleftrightarrow$	$\dfrac{1 - (\alpha \cos \omega_0) z^{-1}}{1 - (2\alpha \cos \omega_0) z^{-1} + \alpha^2 z^{-2}} \quad |z| > |\alpha|$
$\alpha^n \sin(\omega_0 n) \, u[n]$	$\Longleftrightarrow$	$\dfrac{(\alpha \sin \omega_0) z^{-1}}{1 - (2\alpha \cos \omega_0) z^{-1} + \alpha^2 z^{-2}} \quad |z| > |\alpha|$

EXERCISE 5.1.1.   Evaluating the z-Transform Analytically

This problem should be done without the aid of a computer.

(a) Determine the z-transform of each of the following sequences and specify its region of convergence.

(i) $h[n] = \delta[n] + 2\delta[n-1] + 3\delta[n-2]$;

(ii) $s[n] = 4\delta[n-7] + 3\delta[n-9] + 6\delta[n+3]$;

(iii) $g[n] = \left(\frac{1}{3}\right)^n u[n]$;

(iv) $r[n] = \left(\frac{1}{4}\right)^n u[n]$;

(v) $v[n] = g[n] + r[n]$;

(vi) $y[n] = g[n] + r[n] + h[n]$.

(b) For each of the signals above evaluate the z-transform at $z = 0$ and $z = \infty$ to determine whether these points are included in the ROC.

(c) Determine the poles for each sequence that contains poles in the finite z-plane.

EXERCISE 5.1.2. Manipulating Rational z-Transforms

Often a z-transform can be expressed as a rational function of the form $B(z)/A(z)$, which is a ratio of polynomials in z. This exercise provides an introduction to representing and processing rational functions or IIR filters.

Consider the system function $H(z)$ defined as

$$H(z) = \frac{B(z)}{A(z)} = \frac{1 + 2z^{-1} + 3z^{-2} + 4z^{-3} + 5z^{-4}}{1 + 2z^{-1} + 2z^{-2} + 1z^{-3}}$$

(a) This function can be represented as a file using the *create file* option in **x siggen**. Create $H(z)$, type the file to the screen, and record the results.

(b) Create two separate files representing the polynomials $B(z)$ and $A(z)$. In this case, however, use **x convert** to split $H(z)$ into $B(z)$ and $A(z)$. Next use **x rooter** to factor these polynomials. You can see the roots of $B(z)$ and $A(z)$ by typing the resulting files to the screen. In viewing the files, note that the roots appear as the sequence of complex numbers following the file header. These complex numbers are the real and imaginary parts of the roots. The function **x polar** can be used to convert the roots into polar form. Record these roots in terms of their magnitudes and phases.

(c) The roots of the numerator and denominator polynomials provide the essence of the pole/zero plot. Given these roots, the system function can be reconstructed to within a constant multiplier. Starting with the two files containing the poles and zeros of $H(z)$ (in polar form) use **x cartesian** to convert the roots back to their real and imaginary parts, **x rootmult** to expand these roots into polynomials, and **x revert** to merge the numerator and denominator polynomials into one system function file. Record your result. Compare it to $H(z)$ and verify that it is accurate to within a constant gain term.

EXERCISE 5.1.3. Producing Pole/Zero Plots

Pole/zero plots are an important visualization tool in digital signal processing. The plot is formed by factoring the numerator and denominator of a rational system function and plotting the zeros and poles in the z-plane. The function **x polezero** will make pole/zero plots. This exercise demonstrates the use of this function.

(a) Create a file for the system $H(z)$ defined by

$$H(z) = \frac{1 + 3z^{-1} + 3z^{-2} + z^{-3}}{1 + 0.5z^{-1} + 0.3z^{-2} + 0.1z^{-3}}$$

using the *create file* option in **x siggen**.

The program **x polezero**, which is menu driven, will automatically find the roots of the numerator and denominator polynomials and make the plot. Execute the program by typing **x polezero**. Select option (1), *read a file*, in the main menu and enter the name of the file containing the coefficients for $H(z)$. Sketch the pole/zero plot that appears on the screen. A plot of the magnitude response in the range from $-\pi$ to π radians also appears in the lower right part of the screen.

(b) The program will also allow you to list the numerical values of the poles and zeros in terms of their real and imaginary parts. Using this option record the poles and zeros of $H(z)$.

EXERCISE 5.1.4. Evaluating z-Transforms

Find the system function of the LTI systems with the following impulse responses by manipulating the z-transform summation into the form of a geometric series. It is helpful to recognize the following useful identities involving infinite summations:

$$\sum_{n=n_0}^{\infty} a^n = a^{n_0} \sum_{n=0}^{\infty} a^n$$

$$\sum_{n=-\infty}^{-n_0} a^n = \sum_{n=-n_0}^{-\infty} a^n = \sum_{n=n_0}^{\infty} a^{-n}$$

where n_0 is an arbitrary integer.

(a) $h_a[n] = \left(\frac{1}{8}\right)^n u[3-n]$.

(b) $h_b[n] = n \left(\frac{1}{2}\right)^n u[n-2]$.

(c) $h_c[n] = \left(\frac{1}{2}\right)^n u[n-2] + 3^n u[n]$.

Tell whether each of the systems is stable. Note that this exercise does not involve the aid of a computer.

EXERCISE 5.1.5. Calculating System Functions

The systems below represent stable LTI systems that are defined by their impulse responses, denoted $h[n]$.

(i) $h[n] = \left(\frac{1}{5}\right)^n u[n];$

(ii) $h[n] = \left(\frac{1}{5}\right)^n u[n-5];$

(iii) $h[n] = \left(\frac{1}{5}\right)^{n-5} u[n-5];$

(iv) $h[n] = 5^n u[-n-1];$

(v) $h[n] = 2\delta[n] + \left(\frac{1}{5}\right)^n u[n-1].$

(a) Determine (analytically) the system function $H(z)$ for each system.

(b) Sketch the pole/zero plot for each and shade the region of convergence. Do this without the aid of the computer.

(c) For each, construct a data file that represents the system using the *create file* option of **x siggen**. Use **x polezero** to display the poles and zeros of each system function. These plots should agree with your sketches in part (b). Note: **x polezero** will not display poles and zeros at the origin.

EXERCISE 5.1.6. System Functions and Difference Equations

Causal LTI systems are often defined in terms of difference equations relating the input sequence $x[n]$ to the output sequence $y[n]$. The system function can be obtained by evaluating the z-transform of each term in the difference equation using the second property given in Table 5.1, which is called the *shift property*. For example,

$$
\begin{array}{rcl}
x[n] & \leftrightarrow & X(z) \\
x[n-1] & \leftrightarrow & z^{-1}X(z) \\
x[n-2] & \leftrightarrow & z^{-2}X(z) \\
& \vdots & \\
y[n] & \leftrightarrow & Y(z) \\
y[n-1] & \leftrightarrow & z^{-1}Y(z) \\
& \vdots &
\end{array}
$$

The system function $H(z)$ is given by the ratio

$$H(z) = Y(z)/X(z). \tag{5.10}$$

This follows from the last property in Table 5.1, which is known as the *convolution theorem*.

(a) Determine the system functions for the following causal LTI systems and sketch their pole/zero plots.

(i) $y[n] = \frac{1}{2}y[n-1] + x[n];$

(ii) $y[n] = \frac{1}{6}y[n-1] + \frac{1}{8}y[n-2] + x[n] + x[n-1];$

(iii) $y[n] = \displaystyle\sum_{k=-3}^{3} x[n-k].$

This may be done analytically or by using the *create file* option in **x siggen** and **x polezero** to display the plot.

(b) Now consider the causal LTI system with system function

$$H(z) = \frac{1 + z^{-1}}{1 - .5z^{-1} - .1z^{-2}}$$

Determine the linear constant coefficient difference equation (LCCDE) that describes this system.

(c) Using the *create file* option in **x siggen** construct a file containing the coefficients for $H(z)$ in part (b). Using **x filter** and an impulse for the input signal, generate 10 samples of the impulse response of $H(z)$. Remember that the input sequence is the first argument on the command line; $H(z)$ is the second. Now use **x lccde** to generate 10 samples of the impulse response. Compare and record these results.

The system function and the difference equation are two closely related ways to represent an LTI system.

EXERCISE 5.1.7. **The First Backward Difference and the Running Sum**

In Chapter 2 the first backward difference and the running sum operations were defined as

$$y[n] = x[n] - x[n-1]$$

$$y[n] = \sum_{m=-\infty}^{n} x[m]$$

where $x[n]$ is the input to the operator and $y[n]$ is its output. These operations are linear and time invariant and, therefore, can be uniquely characterized by their system functions. This problem examines the system functions of the first backward difference operation and the running sum.

(a) Determine the system function of the first backward difference. Using the *create file* option in **x siggen**, create a file that represents this system function.

(b) Repeat part (a) for the running sum operation. (*Hint:* The running sum can also be expressed as the convolution of $x[n]$ with the step function $u[n]$.)

(c) Use the *sine wave* option in **x siggen** to create 128 points of a cosine wave, $x[n]$. Let the frequency (alpha) equal 0.1, the phase (phi) equal $\pi/2$, and the starting point equal 0. Compute and sketch $v[n]$, the output of the first backward difference system with $x[n]$ as the input. Next compute $y[n]$, the output of the running sum system with input $v[n]$. Use **x filter** to perform the filtering. Remember that the sequence appears as the first argument on the command line and the IIR (running sum) filter appears as the second. Sketch the signals $x[n]$ and $y[n]$ and observe that these operations are inverses of each other.

EXERCISE 5.1.8. **Finding System Functions from Poles and Zeros**

A system may often be described using its poles and zeros. This representation is accurate to within a constant scale factor. If the scale factor and region of convergence are known, then the impulse response is also uniquely specified.

This exercise will make use of the following $H(z)$:

$$H(z) = \frac{(1 + 0.2z^{-1})(1 + 0.3z^{-1})(1 + 0.4z^{-1})(1 + 0.8z^{-1})}{(1 - 0.5z^{-1})(1 - 0.4z^{-1})(1 - 0.3z^{-1})(1 - 0.2z^{-1})}$$

(a) Make a sketch of the pole/zero plot for $H(z)$.

(b) Determine the system function in the form

$$H(z) = \frac{B(z)}{A(z)}$$

and create the corresponding signal file. To do this, first put the roots for the numerator and denominator in separate files using the *create file* option in **x siggen**. Specify the coefficients as complex; the imaginary part will be zero. Use **x rootmult** to multiply these roots together to form $B(z)$ and $A(z)$. Then use the function **x revert** to construct $H(z)$. Record the coefficients for $B(z)$ and $A(z)$.

(c) Use **x filter** to evaluate the impulse response. This can be done by generating an impulse using **x siggen** and using it and $H(z)$ as the inputs to the **x filter** function. Sketch the first 10 points of the result.

EXERCISE 5.1.9. **The DTFT of an IIR System**

The frequency response, $H(e^{j\omega})$, and the system function, $H(z)$, of a stable LTI system are identical when $z = e^{j\omega}$. Previously, we evaluated the frequency response of a filter whose impulse response was of finite length by calculating the discrete-time Fourier transform (DTFT) of the impulse response. The DTFT of a system with an infinite duration impulse response can also be computed in practice if $H(z)$ is a rational function. This can be done by computing the DTFT of the numerator and denominator separately and then dividing. To illustrate this, consider the system function

$$H(z) = \frac{B(z)}{A(z)}$$

where

$$B(z) = 1 + z^{-1} + z^{-2} + z^{-3} + z^{-4} + z^{-5}$$

and

$$A(z) = 1 - 0.9z^{-1}.$$

(a) Create separate files for $B(z)$ and $A(z)$ using the *create file* option of **x siggen**. Display and sketch the DTFT magnitude of $B(z)$ using **x dtft**. The function **x dtft** automatically creates an intermediate DSP file called _dtft_ that contains the frequency-response values that are plotted to the screen. Copy this file to another file called _dtft1_. Now use **x dtft** to display and sketch the frequency response of the denominator polynomial $A(z)$. Since $H(e^{j\omega}) = B(e^{j\omega})/A(e^{j\omega})$ the magnitude response of the system can be obtained simply by dividing the frequency response of the numerator by the frequency response of the denominator. Perform this operation by typing

$$\text{x divide} \qquad \text{\_dtft1\_} \qquad \text{\_dtft\_} \qquad \text{outfile}$$

Use **x look** to display and sketch the magnitude of **outfile** as a continuous waveform.

(b) Now use **x revert** to produce a file for $H(z)$. The function **x dtft** will automatically evaluate the DTFT of a rational function. Display and sketch the DTFT magnitude using **x dtft** directly.

(c) Generate an impulse using **x siggen** and compute 128 points of the impulse response of $H(z)$ using **x filter**. Remember that the first argument of **x filter** should be the impulse and the second should be the IIR filter. Display the impulse response $h[n]$ and observe that the samples of $h[n]$ decay rapidly as n increases. Consequently, the first 128 samples of the impulse response are a good approximation to the true infinite length impulse response. Evaluate the DTFT of the truncated impulse response and sketch the magnitude response.

EXERCISE 5.1.10. **Stability**

If a filter is stable, the ROC of its system function must include the unit circle. If the system is also causal, the poles of the system function must lie inside the unit circle. To illustrate this point, generate files for the following polynomials using the *create file* option of **x siggen**:

(i) $A(z) = 1 + 2z^{-1} + 0.5z^{-2} + 0.2z^{-3}$;

(ii) $B(z) = 1 + 0.5z^{-1} + 0.2z^{-2} + 0.1z^{-3}$;

(iii) $C(z) = 1 + 4z^{-1} + 4z^{-2} + 2z^{-3}$.

Using the function **x revert** construct files for each of the systems given below.

1. $H_1(z) = A(z)/B(z)$

2. $H_2(z) = A(z)/C(z)$

3. $H_3(z) = B(z)/C(z)$

4. $H_4(z) = C(z)/B(z)$

(a) Use **x polezero** to display the pole/zero plots. If each of the systems is causal, determine if each is also stable.

(b) Assume that the system impulse responses are left-sided sequences. In other words, we are assuming that $H_1(z), \ldots, H_4(z)$ are noncausal systems. Tell whether each of the systems is stable.

(c) Now assume that each of the systems has an impulse response that is two-sided, i.e., it extends to infinity in both directions. Which of the systems, if any, is stable? For which system(s) is the answer indeterminate?

EXERCISE 5.1.11. **An Allpole System**

A system with a system function of the form

$$H(z) = 1/A(z)$$

where $A(z)$ is a polynomial in z, is said to be *allpole*. Let $A(z) = 1 + 0.8z^{-1} + 0.4z^{-2} + 0.2z^{-3}$ and assume that the system is causal. Using the *create file* option in **x siggen**, create a file representing $H(z)$ and one representing $A(z)$.

(a) Using **x polezero**, display and sketch the pole/zero and magnitude response plots for $H(z)$ and $A(z)$.

(b) Determine a difference equation with input $x[n]$ and output $v[n]$ that will represent the system, $H(z)$.

(c) Determine another difference equation with input $v[n]$ and output $y[n]$ such that $x[n] = y[n]$. Generate 128 points of a chirp signal using **x siggen** with *alpha1* = 0.05, *alpha2* = 0.001, and *phi* = 1. Use this signal for the input $x[n]$. Apply the function **x filter** to first generate 128 samples of the signal $v[n]$ as defined in part (b). Then use **x filter** to filter $v[n]$ with $A(z)$ to produce 128 samples of $y[n]$. Display $x[n]$ and $y[n]$ one above the other using **x view2** and sketch these signals.

5.2 THE INVERSE z-TRANSFORM

Until now calculating the impulse response of an LTI system has involved exciting the system with a discrete impulse and observing its response. However, the system impulse response can also be found by taking the inverse z-transform of the system function. This section introduces a method for performing this calculation based on the use of partial fractions. This is a relatively easy way to calculate the inverse z-transform.

The discussion that follows will limit itself to the set of rational system functions with distinct poles. These system functions have the form

$$H(z) = \frac{B(z)}{A(z)} = z^{-L}\frac{b_0 + b_1 z^{-1} + b_2 z^{-2} + \cdots + b_M z^{-M}}{1 + a_1 z^{-1} + a_2 z^{-2} + \cdots + a_N z^{-N}} \tag{5.11}$$

where L, M, and N are integers and M and N are assumed to be positive. Note that M and N are the orders of the numerator and denominator polynomials in equation (5.11). $H(z)$, in turn, can also be expressed as

$$H(z) = Cz^{-L}\frac{(1 - \beta_1 z^{-1})(1 - \beta_2 z^{-1})\cdots(1 - \beta_M z^{-1})}{(1 - \alpha_1 z^{-1})(1 - \alpha_2 z^{-1})\cdots(1 - \alpha_N z^{-1})} \qquad (5.12)$$

where the poles $\alpha_1, \alpha_2, \ldots, \alpha_N$ are all assumed to be different. The constant C is a gain and z^{-L} represents an arbitrary delay. System functions of this form can always be expressed as a sum of right-sided components, left-sided components, and finite length components, i.e.,

$$H(z) = z^{-L}\left[\underbrace{\sum_{k=1}^{K-1}\frac{C_k}{1 - \alpha_k z^{-1}}}_{\text{rt-sided terms}} + \underbrace{\sum_{\ell=K}^{N}\frac{C_\ell}{1 - \alpha_\ell z^{-1}}}_{\text{lf-sided terms}} + \underbrace{\sum_{m=0}^{M-N}c_m z^{-m}}_{\text{finite length terms}}\right] \qquad (5.13)$$

where $1 \le K \le N$. Once in this form, the inverse can be found by inspection using equations (5.7–5.9). The challenge is to express $H(z)$ in this form. The method of partial fractions, which is summarized below, accomplishes this.

1. Determine if $H(z)$ is a proper rational function. This means that the condition $M < N$ must be satisfied.[1] When $H(z)$ is proper, there is no finite length component. For example, the function

$$H(z) = \frac{1 + 2z^{-1} + 3z^{-2}}{1 + 3z^{-1} + 4z^{-2}}$$

    is an improper function because both the numerator and denominator have the same order which, in this case, is 2.

2. A partial fraction expansion can only be performed on a proper rational function. If $H(z)$ is not proper, the finite length terms must first be extracted by polynomial division (or *synthetic division* as it is often called). The procedure for dividing polynomials is illustrated later in Example 5.2.

3. Factor the denominator polynomial $A(z)$ into the form

$$A(z) = (1 - \alpha_1 z^{-1})(1 - \alpha_2 z^{-1})\cdots(1 - \alpha_N z^{-1}).$$

4. $H(z)$ can now be written in partial fraction form as

$$H(z) = \frac{C_1}{(1 - \alpha_1 z^{-1})} + \frac{C_2}{(1 - \alpha_2 z^{-1})} + \cdots + \frac{C_N}{(1 - \alpha_N z^{-1})}.$$

---

[1] In terms of equation (5.13) this means that all c_m terms in the last summation are zero. The finite length terms in this equation do not come into play when $M - N < 0$.

The coefficients $\{C_i\}$ can be found by using the formula

$$C_i = \lim_{z \to \alpha_i} H(z)(1 - \alpha_i z^{-1}). \tag{5.14}$$

The coefficients are determined by removing the pole at α_i from the function and evaluating the remaining expression at $z = \alpha_i$. This formula is only valid when the poles are distinct (i.e., all different), although generalizations of it exist for the more general case.

EXAMPLE 5.1.

As an example, let

$$H(z) = \frac{1 + 2z^{-1}}{1 - (1/4)z^{-2}}$$

be the system function for a causal LTI system. Since the order of the numerator is 1 and the order of the denominator is 2, the function is proper and therefore expressible in partial fraction form without any preliminary synthetic division, i.e.,

$$H(z) = \frac{1 + 2z^{-1}}{(1 - \frac{1}{2}z^{-1})(1 + \frac{1}{2}z^{-1})} = \frac{C_1}{(1 - \frac{1}{2}z^{-1})} + \frac{C_2}{(1 + \frac{1}{2}z^{-1})}.$$

The coefficients, C_1 and C_2, can be evaluated using equation (5.14)

$$C_1 = \lim_{z \to 1/2} H(z) \left(1 - \frac{1}{2}z^{-1}\right) = \left. \frac{1 + 2z^{-1}}{1 + (1/2)z^{-1}} \right|_{z=1/2} = \frac{5}{2}.$$

Similarly,

$$C_2 = \left. \frac{1 + 2z^{-1}}{1 - (1/2)z^{-1}} \right|_{z=-1/2} = -\frac{3}{2}.$$

Now that $H(z)$ is in this form, its inverse can be found by inspection. Causality implies that the impulse response is right sided. Therefore, equation (5.7) can be used to give

$$h[n] = \frac{5}{2} \left(\frac{1}{2}\right)^n u[n] - \frac{3}{2} \left(-\frac{1}{2}\right)^n u[n]. \qquad \blacksquare$$

When the order of the numerator is equal to or greater than that of the denominator, a direct application of the partial fraction expansion will not work. The procedure will work, however, if the system function is first transformed into a proper rational function by long division. To illustrate this procedure consider the following example.

EXAMPLE 5.2.

This example will compute the impulse response corresponding to the system function

$$H(z) = \frac{1 + .2z^{-1} + .5z^{-2} + .2z^{-3}}{1 + .5z^{-1}},$$

which again we assume to be causal. The procedure is to remove the polynomial part of the improper function leaving a fractional part that is proper. The division is performed by arranging the numerator and denominator polynomials in descending powers of z^{-1}.

$$
\begin{array}{r}
0.4z^{-2} \quad + \quad 0.2z^{-1} \quad + \quad 3.6 \\[4pt]
\hline
0.5z^{-1} \quad + \quad 1 \;\big|\; 0.2z^{-3} \quad + \quad 0.5z^{-2} \quad + \quad 2.0z^{-1} \quad + \quad 1.0 \\
0.2z^{-3} \quad + \quad 0.4z^{-2} \\[2pt]
\hline
0.1z^{-2} \quad + \quad 2.0z^{-1} \quad + \quad 1.0 \\
0.1z^{-2} \quad + \quad 0.2z^{-1} \\[2pt]
\hline
1.8z^{-1} \quad + \quad 1.0 \\
1.8z^{-1} \quad + \quad 3.6 \\[2pt]
\hline
- \quad 2.6
\end{array}
$$

The division terminates when the remainder is of lower degree than the divisor. In this case, we see that the function can be expressed in the equivalent form

$$
H(z) = 3.6 + 0.2z^{-1} + 0.4z^{-2} + \frac{-2.6}{1 + 0.5z^{-1}}
$$

from which the inverse can be found by inspection.

$$
h[n] = 3.6\delta[n] + 0.2\delta[n-1] + 0.4\delta[n-2] - 2.6(-0.5)^n u[n] \qquad \blacksquare
$$

This inverse transform procedure is explored further in the exercises that follow. Some variations on this approach are introduced along with some alternative methods for finding the inverse z-transform.

EXERCISE 5.2.1. Finding the Inverse z-Transform

Using the preceding examples as illustrations, determine the impulse responses of the causal systems with the following system functions:

(a)

$$
H(z) = \frac{1}{1 - \frac{1}{4}z^{-2}}.
$$

(b)

$$
H(z) = \frac{1 + 2z^{-2}}{1 - \frac{1}{4}z^{-4}}.
$$

This exercise is intended to be worked analytically without the aid of the computer. However, you should feel free to use the computer to verify your answer by checking the first few terms of your impulse responses.

EXERCISE 5.2.2. More Inverse z-Transforms

This exercise is intended to be worked analytically.

(a) Determine the inverse z-transform of the function

$$H(z) = \frac{1 + 2z^{-1} + 2z^{-2} + 2z^{-3}}{1 + 0.5z^{-1}}$$

if the system is causal.

(b) Determine the impulse response of the causal system with system function

$$H(z) = \frac{1 + 2z^{-1} + 0.5z^{-2} + 0.4z^{-3}}{1 + 0.5z^{-1} + 0.1z^{-2}}.$$

The previous discussion suggested long division as a method for converting an improper system function into a proper one plus a polynomial so that the inverse z-transform could be computed. This is not the only possible approach.

Consider the system function

$$H(z) = \frac{1 + 2z^{-2}}{1 - (1/4)z^{-2}}$$

which is an improper rational function that represents a causal system. This can be rewritten in the form

$$H(z) = \frac{1}{1 - (1/4)z^{-2}} + \frac{2z^{-2}}{1 - (1/4)z^{-2}}.$$

The numerator of the second term, $2z^{-2}$, represents a weighted two-sample delay. Thus, if $v[n]$ is the inverse z-transform of the first term, it follows that

$$h[n] = v[n] + 2v[n-2].$$

Note that $v[n]$ can be found without doing any synthetic division. Clearly,

$$V(z) = \frac{1}{(1 - (1/2)z^{-1})(1 + (1/2)z^{-1})}$$

$$= \frac{C_1}{1 - (1/2)z^{-1}} + \frac{C_2}{1 + (1/2)z^{-1}}$$

from which we see

$$v[n] = C_1 \left(\frac{1}{2}\right)^n u[n] + C_2 \left(-\frac{1}{2}\right)^n u[n].$$

EXERCISE 5.2.3. **Completing the Example**

(a) For the example above what are the numerical values of C_1 and C_2?

(b) Determine $h[n]$.

(c) Redo part (b) of Exercise 5.2.2 using this approach and compute the impulse response. Note that the form of the impulse response looks different. Evaluate and record the first three samples of $h[n]$ using both expressions and show that they are in fact the same.

EXERCISE 5.2.4. Computing the Inverse Transform

Consider the system function of the causal LTI system, $H(z)$, where

$$H(z) = \frac{1}{1 - 5z^{-1} + 5.75z^{-2} + 1.25z^{-3} - 1.5z^{-4}}.$$

(a) Determine the impulse response using partial fractions. Use **x siggen** to first create a file for the denominator polynomial and then **x rooter** to factor the polynomial.

(b) Now assume that this system function corresponds to an anticausal system, for which the impulse response is a left-sided sequence. Sketch the pole/zero plot and indicate the region of convergence. Determine the impulse response.

EXERCISE 5.2.5. Effect of the ROC

The z-transform is only unique when the ROC is specified.

(a) Consider the system function

$$H(z) = \frac{1 - 2z^{-2}}{1 + (1/6)z^{-1} - (1/6)z^{-2}}.$$

By considering all possible regions of convergence, determine all possible impulse responses for the system.

(b) Now consider the system function

$$H(z) = \frac{1}{1 - 2z^{-1} + 2z^{-3} - 2z^{-4} + 2z^{-5} - 2z^{-7} + z^{-8}}.$$

Put $H(z)$ into a file using **x siggen**. Use **x polezero** to display the pole/zero plot. Determine the number of impulse response sequences that share this functional form for their z-transform.

EXERCISE 5.2.6. Inversion by Long Division

The approach discussed in this chapter for determining an inverse z-transform involves decomposing the transform into simple terms, each of which can be inverse transformed by inspection. This exercise considers a different method in which the inverse is computed term-by-term using long division.

(a) Consider the system function

$$H(z) = \frac{1}{(1 - (1/2)z^{-1})^2}.$$

Expand the denominator and arrange its terms in an order in which the powers of z^{-1} are ascending. Perform long division and generate five or six terms of the quotient. These are the first few terms of the impulse response. Does this impulse response correspond to a causal system?

(b) Repeat part (a) but reverse the order of the terms in the denominator polynomial so that the powers of z^{-1} are arranged in descending order. Is this system causal?

(c) There are difficulties with this method in terms of finding closed-form expressions for the inverses. Discuss how you would find inverses of transforms with annular ROCs.

The remaining exercises in this section illustrate some other properties of the z-transform.

EXERCISE 5.2.7. **Time Reversal**

This exercise examines the effects of time reversal on the pole/zero plot and the region of convergence.

(a) Create a 6-point chirp sequence, $h[n]$, with *alpha1* = 0.5, *alpha2* = 0.8, *phi* = 1, and starting point zero. Use **x polezero** to display and sketch the pole/zero plot. Now form the sequence $h[-n]$ and sketch the pole/zero plot of the reversed signal. What has happened to the zeros? Show analytically why this relationship is always true.

(b) Consider the sequence $g[n] = (1/3)^n u[n]$, which has a system function with one pole. Based on your observations and analysis in part (a), describe what happens to the pole when $g[n]$ is time reversed. What is the ROC of the system function of the time-reversed signal?

EXERCISE 5.2.8. **Modulation**

This exercise explores the effects of modulating a sequence. Create the sequence

$$x[n] = \begin{cases} (\frac{5}{4})^n, & 0 \le n \le 5 \\ 0, & \text{otherwise} \end{cases}$$

using the exponential option of **x siggen** and plot the zeros using **x polezero**.

(a) Use **x cexp** to generate the following signals:

(i) $x_i[n] = e^{j\frac{\pi}{2}n}x[n]$,

(ii) $x_{ii}[n] = e^{j\pi n}x[n]$,

(iii) $x_{iii}[n] = e^{-j\frac{\pi}{2}n}x[n]$.

Examine their pole/zero plots using **x polezero** and compare these to the pole/zero plot of $x[n]$. Describe the relationships and justify your observations analytically.

(b) Now determine the pole/zero plots of the signals:

    (i) $y_i[n] = \left(\frac{1}{2}\right)^n x[n]$,

    (ii) $y_{ii}[n] = \left(\frac{1}{2}e^{j\pi}\right)^n x[n]$,

    (iii) $y_{iii}[n] = 2^n x[n]$

using **x cexp** and **x polezero**. Note that the modulation variable, *omega*, can be complex, which allows modulation to be performed by damped exponentials. For example, multiplication by $(1/2)^n$ can be performed when $\omega = j\ln(2) = j(0.69315)$. This can be done by specifying 0 0.69315 for *omega* in the **x cexp** function. Compare the resulting pole/zero plots to that of $x[n]$. Examine the roots carefully in polar coordinates and describe their relationship to the roots of $x[n]$. How does the ROC in each case change? Justify your observations with a short proof.

EXERCISE 5.2.9. **Differentiation of the z-Transform**

One of the well-known properties of the z-transform is the derivative property

$$n\,x[n] \longleftrightarrow -z\frac{d}{dz}X(z),$$

which can be derived from the definition. To illustrate this property consider the simple example

$$X(z) = 1 + 2z^{-1} + 3z^{-2} + 4z^{-3}$$

corresponding to

$$x[n] = \delta[n] + 2\delta[n-1] + 3\delta[n-2] + 4\delta[n-3].$$

This exercise does not require the computer.

(a) Analytically compute the expression $-z(d/dz)X(z)$ and take its inverse z-transform. Evaluate the expression $nx[n]$ and compare this with your result.

(b) Using this property, determine the z-transform of $n\alpha^n u[n]$ where $|\alpha| < 1$.

(c) Determine the z-transform of $n^2\alpha^n u[n]$ for $|\alpha| < 1$.

EXERCISE 5.2.10. **Convolution**

This exercise tests your intuition about discrete-time convolution.

(a) Consider the sequences

$$x[n] = 0.75\delta[n] + 0.75\delta[n-1] - 1.25\delta[n-2]$$
$$-1.25\delta[n-3] - 0.75\delta[n-4] - 0.75\delta[n-5]$$
$$+0.25\delta[n-6] + 0.25\delta[n-7]$$

and

$$h[n] = \left(\frac{9}{10}\right)^n u[n].$$

Sketch $x[n]$ and $h[n]$ on paper. Using the method of graphical convolution, produce a rough sketch of $x[n] * h[n]$. Verify your answer by creating a file for $x[n]$ using **x siggen** and a 50-sample approximation of $h[n]$ using the *exponential* option in **x siggen**. Convolve these sequences and sketch the result.

(b) Now repeat part (a) for

$$h[n] = \left(\frac{1}{2}\right)^n u[n].$$

Note that $h[n]$, although infinite in duration, can be expressed as the rational function

$$H(z) = \frac{1}{1 - (1/2)z^{-1}}.$$

Create a file for $H(z)$ and use **x filter** to convolve the signals. Generating 20 samples of the output is sufficient. Why is the output not infinite in length?

5.3 FLOW GRAPHS

We have already seen that a discrete-time LTI system is often specified by a difference equation or by a series of difference equations. The system function of such a system is unique, but the converse is not true. For a given rational system function, it is possible to identify several sets of difference equations that will realize the system. Some of these realizations are better than others. For example, consider the causal LTI system defined by the system function

$$H(z) = 1 + b_1 z^{-1} + b_2 z^{-2} + b_3 z^{-3} + b_4 z^{-4}.$$

One way to implement this system is to use a single difference equation

$$y[n] = x[n] + b_1 x[n-1] + b_2 x[n-2] + b_3 x[n-3] + b_4 x[n-4]$$

where $x[n]$ is the input and $y[n]$ is the output. Another approach is to factor $H(z)$ into the form

$$H(z) = (1 + \alpha_1 z^{-1} + \alpha_2 z^{-2})(1 + \beta_1 z^{-1} + \beta_2 z^{-2})$$

and to express the system function as a pair of difference equations corresponding to a *cascade realization*

$$v[n] = x[n] + \alpha_1 x[n-1] + \alpha_2 x[n-2]$$
$$y[n] = v[n] + \beta_1 v[n-1] + \beta_2 v[n-2].$$

A common realization for a high-order IIR system is as a cascade of second-order sections. The system $H(z)$ is factored into subsystems $H_1(z)$, $H_2(z)$, ..., $H_{N/2}(z)$, each of which has two poles and at most two zeros. The input to the overall system is the input to $H_1(z)$. The system output is the output of $H_{N/2}(z)$, as shown in Fig. 5.4.

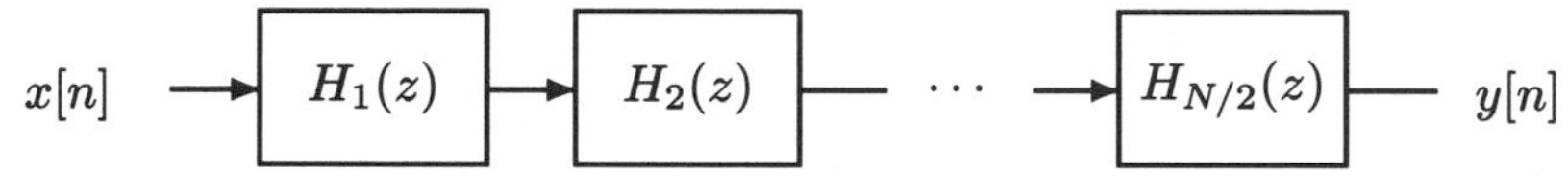

$$x[n] \longrightarrow \boxed{H_1(z)} \longrightarrow \boxed{H_2(z)} \longrightarrow \cdots \longrightarrow \boxed{H_{N/2}(z)} \longrightarrow y[n]$$

Figure 5.4.   A cascade connection of $N/2$ systems.

An alternative is a *parallel implementation,* based on a parallel connection of subsystems or sections as shown in Fig. 5.5 where $H(z) = G_1(z) + G_2(z) + \cdots + G_{N/2}(z)$. These subsystems are typically second-order sections obtained from a partial fraction expansion. Quite often, one of these sections is a constant (i.e., a zero-order subsystem).

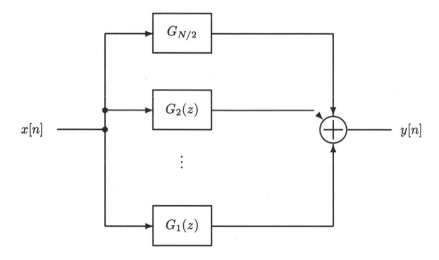

Figure 5.5.   A parallel connection of $N/2$ systems.

Although these different implementations all have the same system function, the performance of the system implemented in these different ways will usually be different due to effects such as coefficient quantization and round-off errors introduced by the limited precision of the numerical processor. In addition, some of these implementations may be faster than others (i.e., they may require fewer arithmetic operations per output sample), while other forms may require less memory.

Digital flow graphs provide a means for defining and manipulating different structures for implementing systems. They can be used to describe a wide variety of systems including nonlinear and time-varying ones, but this text will limit itself to the flow graph elements necessary to represent LTI systems. These basic elements are shown in Fig. 5.6.

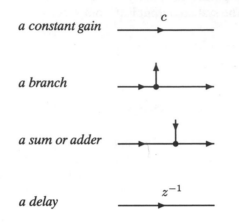

Figure 5.6. The basic elements of a flow graph.

As an example, consider the causal second-order LTI system function

$$H(z) = \frac{b_0 + b_1 z^{-1} + b_2 z^{-2}}{1 + a_1 z^{-1} + a_2 z^{-2}}.$$

A difference equation that will realize it is

$$y[n] = -a_1 y[n-1] - a_2 y[n-2] + b_0 x[n] + b_1 x[n-1] + b_2 x[n-2]. \tag{5.15}$$

The flow graph can be formed by placing the input node $x[n]$ on the left, the output node $y[n]$ on the right, and then using cascades of delay branches to generate the samples $x[n-1]$, $x[n-2]$, $y[n-1]$, and $y[n-2]$. The nodes corresponding to these sample values are then connected together using gains and summers, as shown in Fig. 5.7. The key to drawing flow graphs is to identify the signals represented at the various nodes in the figure and their interrelationships in the difference equation(s). For example, node C in Fig. 5.7 corresponds to $y[n]$. Going through the delay, we see that node F corresponds to $y[n-1]$, i.e., it is $y[n]$ delayed by one sample. Node E is the sum of two branches, one with the signal $-a_1 y[n-1]$, the other with $-a_2 y[n-2]$. Therefore, the signal at node E is $-a_1 y[n-1] - a_2 y[n-2]$. The signal at node A can be found in a similar way. The contributions from the top, middle, and bottom branches are $b_0 x[n]$, $b_1 x[n-1]$, and $b_2 x[n-2]$, respectively. Thus, at node A, the signal is $b_0 x[n] + b_1 x[n-1] + b_2 x[n-2]$. Following this analysis, it is straightforward to verify that the flow graph describes equation (5.15). This kind of analysis allows one to either draw the flow graphs given an equation or determine the difference equation from the flow graph. Since a linear constant coefficient difference equation (LCCDE) can also be expressed as a system function, it is possible to convert any one of these representations to the others.

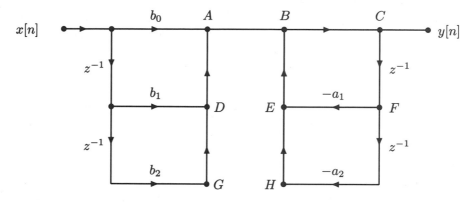

Figure 5.7.   Flowgraph of a second-order IIR system.

For LTI systems, an equivalent system can be formed by taking the transpose of the flowgraph. The transpose is formed by the following three-step process:

1. Interchange the input and output nodes. This corresponds to interchanging $x[n]$ and $y[n]$ in Fig. 5.7, for example.
2. Reverse the directions of the arrows along each branch in the flow graph.
3. Change summing nodes into branch points and branch points into summing nodes.

For LTI systems, the system functions of a flowgraph and of its transpose are the same. This provides a convenient way to construct an alternate but equivalent LTI system. The exercises that follow provide some experience in working with flow graphs and implementing systems.

EXERCISE 5.3.1.   **A Minimum Delay Implementation**

There are many different ways to implement an LTI system, as we have seen. The implementation in Fig. 5.7 is called a *direct form I* implementation. It is straightforward to derive, but it uses more than the minimum number of delays, since past samples of both the input and output need to be stored. This exercise will consider an alternative.

(a)   Referring to Fig. 5.7, split this flow graph into two subsystems: the first with $x[n]$ as its input and the signal at node A as its output, and the second with the signal at node A as its input and $y[n]$ as its output. For convenience, call the first subsystem $H_1(z)$ and the second $H_2(z)$. Thus, the flowgraph is composed of $H_1(z)$ in cascade with $H_2(z)$. Now reverse the order of the systems in the cascade and sketch the resulting flow graph.

(b) Revise this flow graph to remove the redundant delays. The resulting implementation is called a *direct form II* implementation. How many delay elements are required to implement this system?

EXERCISE 5.3.2. Sketching the Flow Graph from LCCDEs

Listed below are several difference equations, each of which describes an LTI system. Sketch the flow graph for each of these systems. (This exercise does not involve the use of the computer.)

(a) $y[n] = x[n] + 3x[n-1] - 2x[n-2] - 0.4y[n-1]$.

(b) $y[n] + 0.2y[n-1] - 0.3y[n-2] + 0.5y[n-3] + 0.6y[n-4] = x[n]$.

(c) $y[n] = x[n] + 2x[n-1] - 3x[n-2] - 2x[n-3] + 0.5x[n-4] + 0.2x[n-5]$.

(d) $y[n] = x[n-4] + 2y[n-3] - 7x[n-7]$.

EXERCISE 5.3.3. Sketching a Flow Graph from a System Function

The system functions shown below correspond to causal LTI systems.

(i) $H(z) = \dfrac{1 + b_1 z^{-1} + b_2 z^{-2} + b_3 z^{-3}}{1 + a_1 z^{-1}}$;

(ii) $H(z) = \dfrac{1 + b_1 z^{-1} + b_2 z^{-2} + b_3 z^{-3}}{1 + a_3 z^{-3}}$;

(iii) $H(z) = c_0 \dfrac{1 + b_2 z^{-2}}{1 + a_1 z^{-1} + a_2 z^{-2}}$.

(a) Sketch the flow graphs for each.

(b) Sketch the transpose of each flow graph in part (a).

(This exercise does not involve the use of the computer.)

EXERCISE 5.3.4. Determining the LCCDE from the Flow Graph

Consider the causal LTI system with input $x[n]$ and output $y[n]$ described by the flow graph shown in Fig. 5.7 where $b_0 = 1$, $b_1 = \frac{3}{4}$, $b_2 = \frac{1}{8}$, $a_1 = \frac{2}{3}$, and $a_2 = \frac{1}{9}$.

(a) Determine the system function, $H(z)$, that corresponds to this system.

(b) Sketch the pole/zero plot associated with the system.

(c) Determine the impulse response of the system.

(d) Sketch the transposed flow graph corresponding to this system.

(This exercise does not involve the use of the computer.)

EXERCISE 5.3.5. **Cascading Second-Order Sections**

Chapter 2 discussed realizations of systems as cascades of second-order subsystems. Such systems are realized by sets of difference equations each of which describes a constituent subsystem.

Consider the causal fourth-order LTI system described by the difference equation

$$y[n] - y[n-1] + \frac{1}{4}y[n-2] + \frac{1}{4}y[n-3] - \frac{1}{8}y[n-4]$$
$$= x[n] + 2x[n-1] + x[n-2].$$

(a) Sketch the flow graph of this system as a direct form I structure.

(b) Find the system function and write it as the product of two second-order system functions. You may wish to use **x siggen** to create a file representing the denominator polynomial of the system function. This polynomial may be decomposed into second-order subsystems using the *delete root* and *write file* options in **x polezero**. Observe that some of the roots are complex. How should the poles within each section be arranged to avoid complex coefficients in the flow graph?

(c) Sketch the flow graph corresponding to the cascade structure that is consistent with your result in part (b).

EXERCISE 5.3.6. **A Parallel Form Implementation**

In Chapter 2, a brief discussion of the parallel form implementation was given. This implementation is based on the representation of a system function as a sum of low-order rational functions, i.e.,

$$H(z) = C_0 + H_1(z) + H_2(z) + \cdots + H_{N/2}(z).$$

Consider the causal fourth-order LTI system given in the previous exercise. Express this system as a sum of second-order subsystems plus a constant and sketch the flow graph corresponding to this realization.

5.4 PLOTTING THE FREQUENCY RESPONSE

Computer-generated plots of the frequency response are routinely produced for systems analysis. These plots can be generated by explicitly evaluating discrete samples of the DTFT using the FFT algorithm that is discussed in the next chapter. These samples can be displayed as a continuous function. While these representations are both accurate and important, there are occasions when a computer is not available and a rough sketch of the frequency response may be satisfactory. Such a rough sketch of the magnitude and phase responses can be obtained from the pole/zero plot using a set of rules based on a geometric interpretation of the DTFT. This method, although not very accurate, allows sketches to be made virtually by inspection. An even more important reason for considering this procedure, however, is the insight that it provides to the system designer.

To develop this method recall that the frequency response is equal to the system function evaluated on the unit circle, i.e.,

$$H(e^{j\omega}) = H(z)|_{z=e^{j\omega}}. \tag{5.16}$$

This allows the frequency response of a system to be related to its poles and zeros. Let $H(z)$ be an arbitrary rational function written in the form we have seen before:

$$H(z) = Cz^{-L} \frac{(1 - \beta_1 z^{-1})(1 - \beta_2 z^{-1}) \cdots (1 - \beta_M z^{-1})}{(1 - \alpha_1 z^{-1})(1 - \alpha_2 z^{-1}) \cdots (1 - \alpha_N z^{-1})}. \tag{5.17}$$

The magnitude of the frequency response is seen to be

$$|H(e^{j\omega})| = |C||e^{-j\omega L}| \frac{|(1 - \beta_1 e^{-j\omega})||(1 - \beta_2 e^{-j\omega})| \cdots |(1 - \beta_M e^{-j\omega})|}{|(1 - \alpha_1 e^{-j\omega})||(1 - \alpha_2 e^{-j\omega})| \cdots |(1 - \alpha_N e^{-j\omega})|}$$

$$= |C| \frac{|(e^{j\omega} - \beta_1)||(e^{j\omega} - \beta_2)| \cdots |(e^{j\omega} - \beta_M)|}{|(e^{j\omega} - \alpha_1)||(e^{j\omega} - \alpha_2)| \cdots |(e^{j\omega} - \alpha_N)|}. \tag{5.18}$$

The last equality follows from the fact that $|e^{j\omega}| = 1$.

A geometrical interpretation can be found by examining the general term $|(e^{j\omega} - \alpha)|$. The point $e^{j\omega}$ lies on the unit circle at an angle ω as shown in Fig. 5.8. The location of the *zero*, which occurs at $z = \alpha = |\alpha|e^{j\theta}$ in the illustration, can also be represented by a vector in the z-plane. It has a magnitude equal to $|\alpha|$ and an angle $\theta = \angle\alpha$. The quantity of interest, however, is $|(e^{j\omega} - \alpha)|$, which is the magnitude of the difference of these two vectors. This is also shown in Fig. 5.8. We observe that as the angle ω corresponding to the point $e^{j\omega}$ on the unit circle is varied, the magnitude of this difference vector, $(|(e^{j\omega} - \alpha)|)$, changes. The implication of this geometric observation is that the frequency-response magnitude can be interpreted as the length of this vector for a given angle ω. Note that the angle ω is simply the frequency in the frequency-response plot and can assume values between 0 and 2π, $-\pi$ and π, or any other interval of 2π radians. For this example the magnitude of the difference vector at $\omega = 0$ is initially large. As ω increases, i.e., moves counterclockwise around the unit circle, the length of the difference vector decreases. It attains its smallest value at $\omega = \pi$ and then increases in length until $\omega = 2\pi$. By plotting this change in length as a function of ω we obtain a plot of the magnitude response for $0 \leq \omega \leq 2\pi$. Notice how the periodicity of the DTFT is illustrated by this geometric viewpoint. If one continues plotting the magnitude response for ever increasing values of ω, we see that this merely corresponds to continually circling the unit circle. Each revolution about the circle represents a period that, in turn, leads to the inherent periodicity property of the DTFT.

The simple example illustrated in Fig. 5.8 can be generalized to represent arbitrary functions of the form of equation (5.18). The magnitude response of such an arbitrary

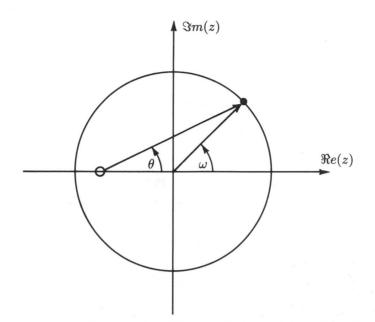

Figure 5.8.   A geometric interpretation of the contribution to the magnitude and phase response due to a single zero.

function can be expressed as the product of individual difference vector magnitudes for the zeros divided by the product of vector magnitudes for the poles. Poles and zeros at the origin can be ignored, since the distance to these is unity for all ω.

Several useful observations arise out of this geometric interpretation. Zeros close to the unit circle cause a dip in the DTFT magnitude in the region near the zero. A zero that lies on the unit circle causes the DTFT magnitude to become zero at that frequency. Similarly, poles close to the unit circle result in peaks in the magnitude response at frequencies close to the angle of the pole. Thus a simple rule for sketching the DTFT magnitude can be formulated. As we move around the unit circle, the DTFT peaks when we pass close to a pole and dips when we pass close to a zero. The sharpness of the peak or dip is directly related to the closeness of the pole or zero to the unit circle. Roots that are far from the unit circle do not affect the magnitude response significantly.

A method for plotting the phase response can be derived similarly, but the procedure is a little more difficult. Consider expressing equation (5.17) in factored form in terms of the individual magnitude and phase terms, i.e.,

$$H(e^{j\omega}) = |A|e^{j\omega K}\ \frac{\displaystyle\prod_{n=0}^{N}|(e^{j\omega} - \beta_n)|e^{j\angle(e^{j\omega} - \beta_n)}}{\displaystyle\prod_{m=0}^{M}|(e^{j\omega} - \alpha_m)|e^{j\angle(e^{j\omega} - \alpha_m)}}. \tag{5.19}$$

The phase is given by

$$\angle H(e^{j\omega}) = \omega K + \sum_{n=0}^{N} \angle(e^{j\omega} - \beta_n) - \sum_{m=0}^{M} \angle(e^{j\omega} - \alpha_m) \qquad (5.20)$$

The contribution of each of the individual phase terms corresponds to the direction (or angle) of the difference vector in Fig. 5.8 from its zero or pole. To compute the magnitude response, we plot the difference vector length as a function of ω. To compute the phase response we plot the angle θ as a function of ω. There is a phase plot associated with each individual pole and zero. To plot the phase response for an arbitrary function of the form of equation (5.19), add the phase plots associated with the zeros and subtract the phase plots associated with the poles. This is the geometric procedure implied by equation (5.20).

EXERCISE 5.4.1. **Sketching the Magnitude Response**

(a) Consider the system with the system function

$$H(z) = 1 - 0.5z^{-1}.$$

Draw a freehand sketch of the pole/zero plot. Divide the unit circle into 16 evenly spaced segments and place a tic mark on the circle at each segment boundary. Measure the distance from the zero to each point on the unit circle. Plot these distances as a function of angle, proceeding counterclockwise around the unit circle. This is a crude plot of the magnitude of the DTFT in the range $0 \le \omega \le 2\pi$. Resketch the plot in the range $-\pi < \omega < \pi$.

Create a file for $H(z)$ using the *create file* option in **x siggen**. Use **x polezero** to check the accuracy of your plot.

(b) Now consider the system with system function

$$H(z) = \frac{1}{1 - 0.5z^{-1}}.$$

Sketch the DTFT magnitude using the same procedure that you used in part (a). Here your plot is the reciprocal of the distances from the pole to points on the unit circle. Again check your answer using **x polezero**.

EXERCISE 5.4.2. **Sketching Magnitude Responses from Pole/Zero Plots**

The geometric interpretation of the pole/zero plot provides a useful method for obtaining a quick sketch of the magnitude response. Consider the following set of system functions expressed in terms of their poles and zeros in polar coordinate form:

System	Poles (mag., angle)	Zeros (mag., angle)
$H_1(z)$	(0.5, 3.1416)	(1.0, 0.0000)
$H_2(z)$	(1.5, 3.1416)	(0.5, 1.5708) (0.5, −1.5708)
$H_3(z)$	(1.0, 0.3927)	(1.0, −0.3927)
$H_4(z)$	(0.6086, 0.96924) (0.6086, −0.96924)	(1.0, 2.4444) (1.0, −2.4444)
$H_5(z)$	(0.6086, 2.17235) (0.6086, −2.17235)	(1.0, 0.6972) (1.0, −0.6972)

Sketch the pole/zero plot for each of these systems. Using the geometric rules developed in the discussion, provide a rough sketch of the magnitude response for each system. Feel free to use the computer to check your results.

EXERCISE 5.4.3. **Sketching the Phase Response**

(a) Consider the system with system function

$$H(z) = 1 - 0.7z^{-1}.$$

Create a file for this function using the *create file* option in **x siggen**. Display the pole/zero plot using **x polezero** and sketch it. Now divide the unit circle into 16 evenly spaced segments and place a tic mark on the circle at each segment boundary. Measure the angle ϕ, which is shown in Fig. 5.8 as a function of ω at each tic mark. Plot these phase angles as a function of ω to obtain a rough phase plot for $0 \le \omega < 2\pi$. To check your answer on the computer, use **x dtft** and display the phase response. *Note:* The phase plot in **x dtft** is in the range $-\pi \le \omega \le \pi$.

(b) Repeat part (a) for the single pole system function

$$H(z) = \frac{1}{1 - 0.7z^{-1}}.$$

Note that for a pole the phase is a plot of $-\phi$ as a function of ω.

EXERCISE 5.4.4. **Poles and Zeros at Infinity**

Poles and zeros at $z = 0$ and $z = \infty$ are often downplayed as being relatively unimportant. This is because these poles and zeros simply cause a shift in the sequence.

This exercise examines the impact of these poles and zeros using examples. It does not require the use of a computer.

Consider the sequence

$$h[n] = \left(\frac{1}{2}\right)^n u[n] + \left(\frac{1}{4}\right)^n u[n]$$

with the *z*-transform

$$H(z) = \frac{2 - (3/4)z^{-1}}{1 - (3/4)z^{-1} + (1/8)z^{-2}}.$$

(a) Sketch the pole/zero plot.

(b) How do the following modifications to the pole/zero plot affect $h[n]$?

    (i)   Introduction of a zero at $z = 0$.

    (ii)   Introduction of a pole at $z = 0$.

    (iii) Introduction of a zero at $z = \infty$.

    (iv) Introduction of a pole at $z = \infty$.

(c) What is the relationship between zeros at $z = 0$ and poles at $z = \infty$?

EXERCISE 5.4.5. **Determining Poles and Zeros from the DTFT**

The impact of poles and zeros (located near the unit circle) on the magnitude response was demonstrated in the introductory discussion. This exercise looks at the reverse relationship by trying to determine the location of poles and zeros from the frequency response.

(a) Using the *create file* option in **x siggen**, create a file to represent the system function of the allpole system

$$H_1(z) = \frac{1}{1 - 0.5z^{-1} + 0.2z^{-2} - 0.1z^{-3} + 0.007z^{-4} + 0.14z^{-5} + 0.15z^{-6}}.$$

This system function has six poles.

    (i)   Use **x dtft** to display and sketch the DTFT magnitude of this allpole filter. Try to sketch the pole/zero plot of this function based on the DTFT magnitude.

    (ii)   Now use **x dtft** to display and sketch the phase response of the system. Based on this plot, attempt a sketch of the pole/zero plot. This one is somewhat more difficult.

    (iii) Use **x polezero** to display the true pole/zero plot and assess the accuracy of your previous attempts.

(b) Consider the system function $H_2(z)$ where

$$H_2(z) = 1 + 1.9z^{-1} + 0.8z^{-2} - 0.8z^{-3} - 0.7z^{-4}.$$

This corresponds to a system with finite length impulse response. Create a file containing $H_2(z)$ using the *create file* option in **x siggen**.

   (i)   Examine and sketch the DTFT magnitude of this function using **x dtft**. Based on this sketch, attempt to draw the pole/zero plot.

   (ii)   Use **x polezero** to display the true pole/zero plot and compare.

(c) What visible cues in the magnitude and phase plots provide information about the pole and zero locations? How are these cues affected by poles and zeros located away from the unit circle?

EXERCISE 5.4.6. Interaction of Poles and Zeros

The poles and zeros of a function have opposite effects on the magnitude response. In this exercise, the interaction of a pole and a zero in close proximity is examined using the following example:

$$H(z) = \frac{1.0 - \alpha z^{-1}}{1.0 - \beta z^{-1}}.$$

(a) Use **x siggen** to create the IIR filter $H(z)$ with $\beta = 0.5$ and $\alpha = 0.8$. Use **x polezero** to display and sketch the pole/zero plot of this function.

(b) Now assume that the pole and zero are variable but constrained to be on the real axis. Use the *change pole/zero* option in **x polezero** to vary the distance between the pole and zero. Pressing the "*d*" key will update the magnitude display. What happens when the pole and zero approach each other? Now move the pole and zero closer to the unit circle. What is the effect of these interactions on the magnitude response?

The DFT and FFT

6

6.1 INTRODUCTION

This chapter will look at the discrete Fourier transform (DFT), and at algorithms for computing it efficiently, particularly the fast Fourier transform (FFT) algorithm. The DFT is a computable transform for sequences of finite length, which can be used to represent the spectral characteristics of a signal. The DFT plays a very important role in digital signal processing (DSP). Thus algorithms for its efficient computation are also important and are treated in this chapter. This study leads naturally to a number of issues related to the computational complexity of signal processing algorithms.

Let $x[n]$ be an N-point complex-valued sequence that is nonzero for values of n between 0 and $N - 1$. The DFT of that sequence, which will be denoted $X[k]$, is the N-point sequence that is defined by the summation

$$X[k] = \sum_{n=0}^{N-1} x[n] W_N^{nk}, \qquad \text{for } k = 0, 1, \dots, N - 1, \tag{6.1}$$

where

$$W_N = e^{-j(2\pi/N)}.$$

The inverse DFT can be evaluated using a similar formula:

$$x[n] = \frac{1}{N} \sum_{k=0}^{N-1} X[k] W_N^{-nk}, \qquad \text{for } n = 0, 1, \dots, N - 1. \tag{6.2}$$

There are at least three interpretations for the DFT, each of which can be helpful in understanding its uses and properties. One point of view regards the DFT as a computable, invertible, linear transformation between two vector spaces—the signal space in which signals are represented as superpositions of delayed unit impulse sequences, and the transform space in which signals are represented as superpositions of complex sinusoids. This interpretation motivates the use of the DFT for waveform coding. A second interpretation regards the DFT as samples of the discrete-time Fourier transform or of the z-transform of the sequence. Comparisons of equation (6.1) with the summations that define the discrete-time Fourier transform (DTFT) and z-transform show that

$$X[k] = X(e^{j\omega})|_{\omega=(2\pi k/N)} = X(z)|_{z=e^{j(2\pi k/N)}}. \tag{6.3}$$

The DFT may be viewed as the z-transform evaluated at the N uniformly spaced points on the unit circle shown in Fig. 6.1. This interpretation motivates the use of the DFT for spectrum analysis. The third interpretation of the DFT regards it as an exact Fourier series representation for the periodic extension of $x[n]$ with period N. The periodic extension of $x[n]$ is formed by repeating $x[n]$ every N samples in both the positive and negative directions. This third interpretation is the most useful for understanding many of the properties of $X[k]$. These are briefly summarized in the next section.

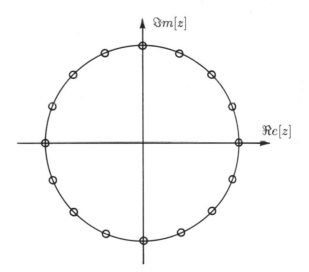

Figure 6.1.   The z-plane locations of the samples of a 16-point DFT.

EXERCISE 6.1.1. **Some Simple DFT Properties**

The programs **x dft** and **x idft** are unsophisticated programs for evaluating the forward and inverse discrete Fourier transforms of an N-point sequence. They evaluate the DFT and IDFT using the formulas given in equations (6.1) and (6.2) for any positive integer value of N. In this exercise, you examine some simple properties of the DFT, first experimentally by computing specific DFTs and then analytically. In working the various parts of this exercise, you will be asked to consider arbitrary sequences that satisfy certain conditions. You may wish to repeat the experiments for several different sequences that satisfy the conditions to obtain a better illustration of the particular property.

(a) Consider a 16-point *circularly antisymmetric* sequence, $x_a[n]$. This is a sequence that satisfies the relation

$$x_a[n] = -x_a[15 - n], \qquad n = 0, 1, \ldots, 15.$$

Use the function **x fdesign** to design a filter of this type. Specifically, specify a *highpass* filter with a cutoff frequency of 0.3 radian. Using **x view**, display the signal and provide a sketch.

Next consider a 16-point sequence, $x_s[n]$, that satisfies the relation

$$x_s[n] = x_s[15 - n], \qquad n = 0, 1, \ldots, 15.$$

Such a sequence is said to be *circularly symmetric*. Use **x mag** to form the sequence $x_s[n] = |x_a[n]|$. Again use **x view** to display the signal. Provide a sketch.

The DFTs of circularly symmetric and antisymmetric signals exhibit a characteristic symmetry. Use **x dft** to compute $X_a[k]$ and $X_s[k]$. Display the real and imaginary parts of these transforms using **x view2**. Clearly describe the characteristics of the DFTs for the circularly symmetric and antisymmetric cases. Feel free to try other test examples before drawing your conclusions.

(b) Compute the *sum* of $x_a[n]$ and $x_s[n]$ that you generated in part (a) using **x add**. Display and sketch the real and imaginary parts of its DFT using **x dft** and **x view**. How are these parts related to those of the two earlier DFTs?

(c) Consider a 16-point "periodic" sequence, $x_p[n]$, that satisfies the relation

$$x_p[n] = x_p[n + 4] = x_p[n + 8] = x_p[n + 12], \qquad n = 0, 1, 2, 3.$$

Use **x siggen** to design a 4-point ramp function with a slope of one and starting point zero. Next use **x lshift** and **x add** in succession to produce $x_p[n]$. Use **x view** to display $x_p[n]$ and verify that it is correct. Compute the DFT and display its real and imaginary parts. Describe the structure imposed on the DFT by the periodic nature of $x_p[n]$. Provide sketches of the DFTs.

(d) Consider a 16-point sequence, $x[n]$, that is zero for its odd-indexed samples, i.e., $x[n] = 0$ for $n = 1, 3, \ldots, 15$. Design an 8-point random signal using **x rgen**. Then use **x upsample** with $M = 2$. This results in a 15-point sequence. Use **x zeropad** with ending value 15 to append the final zero. This will produce $x[n]$. Display $x[n]$ with **x view** to verify that it is correct. Compute and sketch the DFT of $x[n]$ and describe the structure you observe.

(e) Generate the 16-point *sinusoid*

$$x[n] = \cos\left(\frac{5\pi}{8}n\right), \qquad n = 0, 1, \ldots, 15$$

using **x siggen**. Compute the DFT and display its real and imaginary parts. Describe the special nature of $X[k]$. Include a sketch of $X[k]$ with its amplitude scale carefully labeled.

(f) Each of these examples illustrates a property of the DFT. In each case, state the property and show analytically, using the definition of the DFT, that it is true.

EXERCISE 6.1.2. Direct DFT Evaluation

The DFT definition in equation (6.1) suggests that the total number of multiplications and additions should be proportional to N^2, since each sample of $X[k]$ requires about N complex multiplications and additions and N samples of $X(k)$ need to be computed. This quadratic relationship between the DFT length and the amount of computation makes the direct evaluation of large DFTs computationally unattractive. To illustrate this quadratic dependency use the **x dft** function to evaluate the discrete Fourier transforms of several arbitrary sequences of lengths 32, 64, 128, and 256. The function **x rgen** may be used to create the sequences. If your machine takes longer than five or ten minutes to evaluate the longer DFTs, you may wish to terminate the program prematurely. If, on the other hand, your machine is very fast and evaluates all of the DTFs in only a few seconds, increase the transform length appropriately to obtain useful measurements.

Record the computation times required for each case. Observe that as the DFT size N is increased linearly, the computation time increases at a disproportionate rate. The data you record may not show an exact quadratic relationship because **x dft** must also perform I/O operations that do not have a quadratic time dependence with respect to sequence size.

EXERCISE 6.1.3. Zero Padding

The DFT of an N-point complex sequence is an N-point complex sequence in general. There are occasions, however, when the length of the DFT should be longer than the length of the sequence. This will be true, for example, when the DFTs of two finite length sequences of different lengths need to be added together. The length of the DFT of the shorter sequence needs to be increased before adding. Another example would be in the display of a high-resolution spectrum of a short sequence. There a large number of spectral or DFT samples is required. In this exercise, we consider how to expand the length of the DFT without distorting the general properties of the transform.

(a) Design a 16-point lowpass filter, $h[n]$, with a cutoff frequency of $\pi/2$ radians using the function **x fdesign**. Evaluate and sketch its magnitude response using **x dtft**. Next evaluate and sketch the magnitude of the 16-point DFT of $h[n]$ using

x dft and **x view**. Since the DFT is a sampled version of the DTFT, the appearance of $|H(k)|$ is known. However, there is an issue associated with the frequency interval being displayed in the output. The function **x dtft** displays the frequency response in the interval $-\pi \le \omega \le \pi$. Examine the definition of the DFT as well as your plot and determine the frequency interval in which the DFT samples occur.

(b) Now embed this sequence $h[n]$ into an $M = 64$-point sequence, $y[n]$, defined as

$$y[n] = \begin{cases} h[n], & n = 0, 1, \ldots, 15 \\ 0, & \text{otherwise.} \end{cases}$$

This operation is called *zero padding*. It can be implemented using the function **x zeropad** where the ending point should be specified to be 63. Evaluate the 64-point DFT of $y[n]$. Since the direct implementation of the DFT is time-consuming for long sequences, you may wish to use the equivalent, but more efficient realization **x fft** in place of **x dft**. Use **x view** to display and sketch the magnitude of $Y[k]$. How is $Y[k]$ related to $H[k]$?

(c) Repeat part (b) for $M = 256$. In this case the ending point parameter in **x zeropad** will be 255.

(d) Relate the values of the M-point DFT, $Y[k]$, to the discrete-time Fourier transform and z-transform of $h[n]$ following equation (6.3). Using the idea of zero padding, write a macro to display the DTFT magnitude of $h[n]$ that essentially mimics the function **x dtft**. You may use the **x look** function to display your output as a continuous plot. Test the operation of your macro on $h[n]$ and sketch the result.

EXERCISE 6.1.4. **Spatial Aliasing**

It is occasionally desirable to have the number of frequency-domain samples be less than the original number of time samples. For example, we may have a very long sequence and wish to obtain only a small number of evenly spaced samples in the frequency domain. This case is the complement of the case addressed in the previous exercise. Here we are interested in evaluating M samples of the Fourier transform of an N-point sequence when $N > M$.

(a) Using **x siggen**, generate the exponential

$$x[n] = \alpha^n, \qquad n = 0, 1, \ldots, 127$$

where $\alpha = \frac{39}{40}$. Display and sketch the DFT magnitude of this sequence using **x fft** and **x view**.

(b) Now generate a new 64-point sequence $y[n]$ by *spatial aliasing*. This is defined by

$$y[n] = x[n] + x[n + 64], \qquad n = 0, 1, \ldots, 63$$

and can be created using **x lshift** and **x add**. The result of shifting and adding is a sequence beginning at -64 instead of zero. Use **x extract** to extract the samples from 0 to 63. Evaluate the DFT magnitude of $y[n]$ and compare it with the result in (a).

(c) Again consider the *spatial aliasing* concept on the sequence $x[n]$ defined in part (a). Outline the steps for computing the 32-point DFT of this sequence. Compute the 32-point DFT in this manner and sketch the DFT magnitude.

6.2 PROPERTIES OF THE DFT

Many properties of the DFT can be exploited to reduce computation in the design of systems. Some of these are similar to properties of the discrete-time Fourier transform and of the z-transform. Table 6.1 contains an abbreviated list of these properties. In this table the notation $x[[n]]_N$ refers to the N-point periodic extension of $x[n]$ evaluated over the interval $0, 1, \ldots, N-1$. In other words, $x[[n]]_N = x[n]$ in the interval $0, 1, \ldots, N-1$. For n outside of this interval, $x[[n]]_N = x[n + kN]$ where k is the integer (negative or positive) such that $n + kN$ lies within the interval.

Table 6.1.   Some Properties of the DFT[a]

	N-**Point Sequence**	N-**Point DFT**
1.	$x[n]$	$X[k]$
2.	$y[n]$	$Y[k]$
3.	$ax[n] + by[n]$	$aX[k] + bY[k]$
4.	$x[[n + n_0]]_N$	$W_N^{-kn_0} X[k]$
5.	$W_N^{nk_0} x[n]$	$X[[k - k_0]]_N$
6.	$\displaystyle\sum_{m=0}^{N-1} x[m]y[[n - m]]_N$	$X[k]Y[k]$
7.	$x[n]y[n]$	$\dfrac{1}{N}\displaystyle\sum_{\ell=0}^{N-1} X[\ell]Y[[k - \ell]]_N$
8.	$x^*[n]$	$X^*[[-k]]_N$

[a]The N-point signals and N-point DFTs are always assumed to be sequences beginning at 0 and ending at $N - 1$.

Since the DFT is an N-point to N-point transformation, only the points in the region $0 \le n \le N - 1$ or $0 \le k \le N - 1$ are important. Thus a sequence of this form can always be multiplied by $r_N[n]$ where

$$r_N[n] = \begin{cases} 1, & n = 0, 1, \ldots, N-1 \\ 0, & \text{otherwise} \end{cases}$$

without changing the value of the N-point sequence. Rectangular windows are often used to delineate the index range.

The exercises in this section were chosen to help understand a number of these properties.

EXERCISE 6.2.1. **Introduction to Circular Shifting**

The DFT and DTFT have many similar properties. The major difference, however, is that operations for the DFT are assumed to be circular. The DFT is an N-point to N-point transformation. The input sequence is always considered over the range $0 \le n \le N-1$, and the transform is always defined for $0 \le k \le N-1$. Consequently, operations such as shifting (which inherently translates the sequence to a different interval) must be interpreted differently here.

In this exercise, we investigate the operation of performing circular shifts in the time and DFT domains.

(a) Generate the 16-point sequence

$$x[n] = \begin{cases} 1, & n = 0 \\ 0, & n = 1, 2, \ldots, 15 \end{cases}$$

using the *square wave* option in **x siggen**. Display and sketch $x[n]$ using **x view**. Use the circular shifting function **x cshift** to shift $x[n]$ by 5, 10, 15, 17, and 20. Display and sketch these shifted signals. These circular shifts can be expressed mathematically as $x[[n - n_0]]_N$, where in this case $N = 16$. When n_0 is positive, we have a circular right shift. For n_0 negative, a circular left shift results. Try sketching $x[[n + 6]]_N$ without the aid of the computer. Now check your answer.

(b) Circular shifts can be applied to any N-point sequence regardless of whether it is a time-domain sequence like $x[n]$ or a transformed sequence like $X[k]$. Let $v[n] = x[[n - 18]]_N$. Compute and display the real and imaginary parts of $V[k]$ and $V[[k + 2]]_N$ using **x dft**, **x cshift**, and **x view2**. Sketch the results.

EXERCISE 6.2.2. **Circular Convolution**

The *circular convolution property* (property 6 in Table 6.1) states that the inverse DFT of the product of two DFTs is the circular convolution of the two sequences. The circular convolution depends on the value of the parameter, N.

(a) Generate the 10-point sequence $x[n]$ where

$$x[n] = \begin{cases} 1, & n = 0, 1, \ldots, 7 \\ 0, & n = 8, 9. \end{cases}$$

Form the 10-point circular convolution of $x[n]$ with itself using the function **x cconvolve**. Display and sketch the result using **x view**.

(b) Now compute this circular convolution by computing the 10-point DFT of $x[n]$ using **x dft**, squaring the result using **x multiply**, and computing an inverse DFT using **x idft**. Again display and sketch the output and compare it with your results in part (a). Observe that your output is complex. Explain why the imaginary part is not exactly zero as you would expect.

EXERCISE 6.2.3.   Circular and Linear Convolutions

The N-point sequence length constraint implied by the DFT prohibits operations that expand the sequence length beyond N-points. Linear convolution has the property that the convolution of two N-point sequences is $2N - 1$ points long. Circular convolution, by contrast, produces a sequence of length N and is defined as

$$x[n] \circledast_N h[n] = \left(\sum_{m=0}^{N-1} x[[n - m]]_N \, h[m] \right) r_N[n]$$

where $r_N[n]$ is the rectangular window

$$r_N[n] - \begin{cases} 1, & n = 0, 1, 2, \ldots, N - 1 \\ 0, & \text{otherwise.} \end{cases}$$

(a) Use **x siggen** to generate a 16-point ramp function beginning at $n = 0$. Now use **x cconvolve** to form the 16-point circular convolution

$$y_1[n] = x[n] \circledast_{16} x[n].$$

Display and sketch the output using **x view**.

(b) Use **x convolve** to perform the linear convolution

$$y_2[n] = x[n] * x[n].$$

Use **x view2** to display and sketch $y_1[n]$ and $y_2[n]$ and observe that they appear to be quite different.

(c) Now take $y_2[n]$ and spatially alias it with itself, by forming the sequence

$$y_3[n] = \begin{cases} y_2[n] + y_2[n + 16], & n = 0, 1, 2, \ldots, 15 \\ 0, & \text{otherwise.} \end{cases}$$

Compare the result with $y_1[n]$. *Note:* you may use **x lshift** with a shift of -16 to perform the shifting and the **x extract** function to extract the samples between 0 to 15. Observe that circular convolution can be interpreted in terms of linear convolution.

EXERCISE 6.2.4. **Linear Convolutions via Circular Convolutions**

(a) Create a 64-point square wave, $x[n]$, with pulse length 5 and period 8 starting at $n = 0$ using **x siggen**. Then create the signal $h[n] = u[n] - u[n - 16]$ using the block option in **x siggen**. Find the *linear* convolution of these two signals using **x convolve** and use **x view** to display it. Give a rough sketch of the result.

(b) Extend $h[n]$ to 64 points using **x zeropad** and form the 64-point circular convolution of $x[n]$ with $h[n]$ using **x cconvolve**. Provide a rough sketch of the result.

(c) Now use **x zeropad** to extend each of the sequences to 128 points. Evaluate the convolution as before, using N-point circular convolution with $N = 128$. How does your result compare with your answers in (a) and (b) above? What is the smallest value of N that will make the circular convolution equal to the linear convolution for all values of n between 0 and $N - 1$? Check your answer using the computer.

EXERCISE 6.2.5. **Circular Shift**

Property 4 in Table 6.1 is known as the *circular-shift property*.

(a) The sequence $y[n] = x[[n + n_0]]_N r_N[n]$ is a (left) circular shift of $x[n]$ of n_0 samples where $r_N[n]$ was defined in Exercise 6.2.3. Use **x siggen** to generate a 15-point triangular sequence, $x[n]$. Then generate another 15-point sequence, $y[n] = x[[n - 3]]_N r_N[n]$, using **x cshift**.

(b) Evaluate the two 15-point DFTs, $X[k]$ and $Y[k]$, using **x dft**. Display and sketch the magnitudes of the two DFTs using **x view2**.

(c) The circular shift can be achieved by circularly convolving $x[n]$ with a particular sequence $h[n]$. Determine the sequence $h[n]$ and sketch the magnitude and phase of its DFT.

(d) Compute and record the inverse 15-point DFT of $Y[k]/X[k]$ using **x divide**, **x idft**, and **x view**, and compare it with your result in (c). Summarize your observations.

EXERCISE 6.2.6. **Circular Modulation**

When a sequence is multiplied by a complex exponential, the DFT of the sequence is circularly shifted. This is called the *circular-modulation property*, and it is the dual of the circular-shift property.

(a) Use **x siggen** to generate the 64-point sequence

$$x[n] = \begin{cases} 1, & 0 \le n \le 15 \\ 0, & 16 \le n \le 64. \end{cases}$$

Plot and sketch the DFT magnitude of this 64-point sequence using the function **x fft** and **x view**.

(b) Evaluate and plot the DFTs of the three related sequences

  (i) $(-1)^n x[n]$;

  (ii) $(j)^n x[n]$;

  (iii) $(-j)^n x[n]$.

(This modulation can be performed using **x cexp** by specifying $\omega_0 = \pi, (\pi/2)$ and $-(\pi/2)$, respectively.) How are these related to $X[k]$?

6.3 UNDERSTANDING THE FFT

Because the DFT is important in signal processing, algorithms for its evaluation have been actively studied for more than thirty years, and a number of highly efficient algorithms for calculating the DFT have been discovered. Collectively these are known as fast Fourier transform (FFT) algorithms. They exploit the periodicity and the structure of the kernels W_N^{nk} for various values of N. FFT algorithms are of two types— Cooley–Tukey algorithms, which have been known to the signal processing community since 1965, and prime factor algorithms, which were discovered by Good in 1959, and more fully developed in the 1970s. The former can be more easily programmed, particularly for cases where the length, N, is variable, but the latter algorithms are more efficient.

Cooley–Tukey algorithms require that the DFT length, N, be composite (i.e., factorable). Therefore, let $N = N_1 N_2$ for integer values of N_1 and N_2 and define

$$n = N_1 n_2 + n_1, \quad n_1 = 0, 1, \ldots, N_1 - 1; \quad n_2 = 0, 1, \ldots, N_2 - 1$$
$$k = k_2 + N_2 k_1, \quad k_1 = 0, 1, \ldots, N_1 - 1; \quad k_2 = 0, 1, \ldots, N_2 - 1.$$

Then the DFT sum can be written as:

$$X[k_2 + N_2 k_1] = \sum_{n_1=0}^{N_1-1} \sum_{n_2=0}^{N_2-1} x[N_1 n_2 + n_1] W_N^{(N_1 n_2 + n_1)(k_2 + N_2 k_1)}$$

$$= \sum_{n_1=0}^{N_1-1} W_N^{N_2 n_1 k_1} \left\{ W_N^{n_1 k_2} \sum_{n_2=0}^{N_2-1} x[N_1 n_2 + n_1] W_N^{N_1 n_2 k_2} \right\}$$

$$= \sum_{n_1=0}^{N_1-1} W_{N_1}^{n_1 k_1} \left\{ W_N^{n_1 k_2} \sum_{n_2=0}^{N_2-1} x[N_1 n_2 + n_1] W_{N_2}^{n_2 k_2} \right\},$$

for $k_1 = 0, 1, \ldots, N_1 - 1$, $k_2 = 0, 1, \ldots, N_2 - 1$. (Observe in the process of expanding the kernel $W_N^{(N_1 n_2 + n_1)(k_2 + N_2 k_1)}$ into the separate terms $W_N^{N_2 n_1 k_1}$, $W_N^{n_1 k_2}$, $W_N^{N_1 n_2 k_2}$, and $W_N^{N_1 N_2 n_2 k_1}$ that $W_N^{N_1 N_2 n_2 k_1} = e^{j2\pi n_2 k_1} = 1$. In addition, note that

$W_N^{N_2 n_1 k_1} = W_{N_1}^{n_1 k_1}$ and $W_N^{N_1 n_2 k_2} = W_{N_2}^{n_2 k_2}$.) The inner summation represents an N_2-point DFT for each value of n_1, and the summation outside the braces represents an additional N_1-point DFT for each value of k_2. This decomposition has thus succeeded in breaking down the N-point DFT into several N_1-point and N_2-point DFTs. If C_N represents the number of complex multiplications required by the algorithm to evaluate the N-point DFT, then

$$C_N = N_2 C_{N_1} + N_1 C_{N_2} + N.$$

The last term in this expression is due to the *twiddle factor* multiplications by $W_N^{n_1 k_2}$. When either N_1 or N_2 is composite, the same decomposition can be used to evaluate the smaller DFTs and this procedure can be continued until the lengths of all of the individual DFTs are prime.

In general, efficient FFT algorithms rely on N being a highly composite integer. N-point FFTs of this type, which exploit composite integers of the form $N = R^\nu$, are called *radix-R* algorithms, the most popular being radix-2 and radix-4 algorithms. For such composite integers the number of complex multiplies required to evaluate an N-point DFT is approximately $(N/2) \log_2 N$. Since one complex multiply requires four real multiplies and two real adds, this is equivalent to $2N \log_2 N$ real multiplies. The total number of real adds required is approximately $3N \log_2 N$. If the FFT is based on factorizations by factors other than powers of R, it is called a *mixed-radix* algorithm.

This method of decomposing the input sequence into smaller subsequences can be viewed as *decimation* of the time signal. The smaller DFTs are then performed on these decimated sequences. This property is the motivation for the name *decimation-in-time* or *DIT* FFT algorithm. The flow graph of an 8-point radix-2 decimation-in-time Cooley–Tukey algorithm is shown in Fig. 6.2. The transpose of the DIT flow graph results in an alternate or dual form of this FFT algorithm called *decimation-in-frequency* or *DIF*.

The prime factor algorithms are similar to the extent that a large DFT evaluation is broken down into a number of small ones; the algorithms differ, however, in their details. The prime factor algorithms also require that N be composite; more specifically, they require that $N = N_1 N_2$, where N_1 and N_2 are relatively prime. They exploit the substitutions

$$n = [[\mu_1 n_1 + \mu_2 n_2]]_N$$
$$k = [[\mu_1 k_1 + \mu_2 k_2]]_N.$$

Results from number theory (specifically the Chinese remainder theorem) have established that the numbers μ_1 and μ_2 exist such that values of n_1 in the range between 0 and $N_1 - 1$, and n_2 in the range between 0 and $N_2 - 1$ will map to unique values of n between 0 and $N_1 N_2 - 1$. With this substitution and some simplification of the resulting expression (aided by some additional results from number theory) the DFT summation becomes

$$X[[\mu_1 k_1 + \mu_2 k_2]]_N = \sum_{n_1=0}^{N_1-1} \sum_{n_2=0}^{N_2-1} x[[\mu_1 n_1 + \mu_2 n_2]]_N W_{N_1}^{R_1 n_1 k_1} W_{N_2}^{R_2 n_2 k_2}.$$

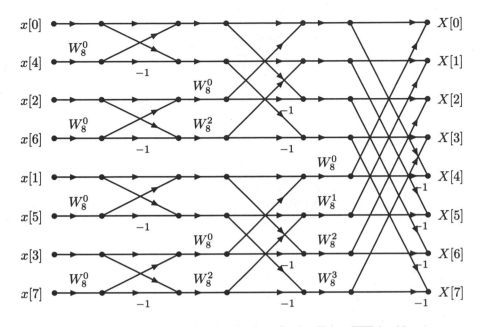

Figure 6.2. A radix-2 decimation-in-time Cooley–Tukey FFT for $N = 8$.

Defining

$$m_1 = [[R_1 n_1]]_{N_1}$$
$$m_2 = [[R_2 n_2]]_{N_2}$$

it is seen that the DFT can be evaluated by a rearrangement of the input sequence followed by N_1 DFTs of length N_2 points taken with respect to the variable m_2. These, in turn, are followed by N_2 N_1-point DFTs taken with respect to m_1. The total computation is

$$C_N = N_2 C_{N_1} + N_1 C_{N_2}.$$

This is less than required by the Cooley–Tukey algorithms because there are no twiddle factor multiplications. The requirement that N_1 and N_2 be relatively prime, however, complicates the programming of a general subroutine that will work for many values of N.

The exercises in this section are intended to help you understand the FFT algorithm and some of its relatives. In many cases this involves writing macros and running them. In this computing environment running macros is inefficient and the computational savings that are made possible by many of the algorithmic tricks may not be reflected in reductions in clock times. To get real savings in time requires explicitly writing new programs and compiling and running them.

EXERCISE 6.3.1. **Radix-2 FFT**

In a radix-2 decimation-in-time FFT, $N = 2^\nu$ and, in the terminology of this section, $N_1 = 2$ and $N_2 = N/2$. This leads to

$$X[k] = G[k] + W_N^k H[k], \qquad k = 0, 1, \ldots, N-1$$

where $G[k]$ is formed by computing the $N/2$-point DFT of the even samples of $x[n]$ and $H[k]$ is formed by computing the $N/2$-point DFT of the odd samples. The two smaller DFTs are extended to N samples by periodic extension before this formula is applied. The intent of this exercise is to verify this formula.

(a) Use the *square wave* option in **x siggen** to produce a 64-point signal consisting of 9 ones followed by 55 zeros and starting at $n = 0$. This signal will be $x[n]$. Compute its 64-point DFT, $X[k]$, using **x fft**. Use **x view** to display and sketch its real and imaginary parts.

(b) Write a macro to calculate $X[k]$ using the formula above. One scheme for doing this would involve the following steps:

   (i) Use **x dnsample** and **x lshift** to generate the 32-point sequences $g[n]$ and $h[n]$.

   (ii) Use **x fft** to calculate the 32-point DFTs, $G[k]$ and $H[k]$.

   (iii) Use **x lshift** and **x add** to periodically extend $G[k]$ and $H[k]$ to 64-point sequences. In other words, let

$$G[k] \Rightarrow G[k] + G[k - 32]$$
$$H[k] \Rightarrow H[k] + H[k - 32].$$

   (iv) Then use **x cexp** to perform the $W_N^k H(k)$ operation prior to adding the sequences.

(c) Use your macro to evaluate the DFT of $x[n]$ from part (a). Plot your result with the result from part (a) using **x view2** and compare them. Are they identical?

EXERCISE 6.3.2. **Inverse FFT**

The evaluation of an inverse DFT is obviously very similar to the evaluation of the DFT itself. This exercise will consider means for using FFT programs to calculate inverse DFTs. There are at least three ways by which this can be done.

Method I: Modify the program itself. In the inverse DFT, W_N is replaced by its complex conjugate and the result is divided by N.

Method II: Use the fact that $x^*[n] = \text{DFT}\{X^*[k]\}/N$ to evaluate the inverse DFT using a forward DFT program. This means that you should conjugate $X[k]$, pass the result through a (forward) DFT subroutine, conjugate the result, and divide it by N.

Method III: Use the fact that $x[[N-n]]_N = \text{DFT}\{X[k]\}/N$. This means that you should pass $X[k]$ through a DFT subroutine, (circularly) reverse the order of the result, and divide it by N.

(a) Write a macro that will evaluate an inverse DFT of an 8-point sequence using the procedure outlined in Method II above. The macro should contain the functions **x conjugate, x fft**, and **x gain**.

(b) Write a macro for the inverse DFT of an 8-point sequence based on Method III. Carefully consider the operations implied by $x[N-n]$. It will involve the use of **x reverse, x lshift**, and **x cshift.**

(c) Use **x siggen** to create an 8-point ramp sequence with starting point zero. Compute the inverse DFT of the ramp using the macros in parts (a) and (b). Display their real and imaginary parts using **x view2**. Compute the inverse directly using **x ifft** and verify that your results are correct. Sketch the inverse DFT of the ramp.

EXERCISE 6.3.3. Computational Complexity of the FFT

In this exercise, you will measure the execution times of the FFT algorithm for different length sequences to assess the arithmetic complexity of the algorithm as a function of length. Plot the FFT execution times of arbitrary random sequences with lengths that are a power of 2 on a time scale from zero to about five minutes. Start by creating arbitrary sequences of lengths 16, 64, 256, 1024, and 2048 using **x rgen**. Evaluate the FFT using **x fft** and record the computation times. If the execution times are very short due to the high speed of your computer, use longer length sequences. If, on the other hand, execution times are very long, choose smaller lengths that result in execution times of no more than a few minutes.

Note that for the shorter sequences the I/O overhead of the program will dominate the computation time of the FFT itself. How well do your measurements fit the complexity formula (which is proportional to $N \log_2 N$) based on counting multiplications and additions? Compare the run times with those of the straightforward function **x dft** on those sequences for which **x dft** can run in less than 5 minutes.

EXERCISE 6.3.4. Decimation-in-Frequency (DIF) FFT

The radix-2 decimation-in-time (DIT) DFT is based on the substitutions

$$n = N_1 n_2 + n_1$$
$$k = k_2 + N_2 k_1$$

with $N_1 = 2$ and $N_2 = N/2$ at the first stage. Decimation-in-frequency or DIF algorithms result by reversing the roles of N_1 and N_2. The flowchart of a DIF algorithm is the transpose of that of a DIT. This leads to the following basic decomposition for the decimation-in-frequency FFT:

$$g[n] = (x[n] + x[n + N/2])$$
$$h[n] = (x[n] - x[n + N/2]) \cdot W_N^n$$

and

$$X[2k] = G[k] \qquad (6.4)$$

$$X[2k + 1] = H[k]. \qquad (6.5)$$

The DFT of $g[n]$ yields the even samples of $X[k]$, and the DFT of $h[n]$ yields its odd samples. The same decomposition can be used to compute the $N/2$-point DFTs of $g[n]$ and $h[n]$ to achieve further efficiency.

(a) Draw the flowchart of a complete 8-point DIF FFT.

(b) Using Exercise 6.3.1 as a guide, write a macro to calculate $X(k)$ using equations (6.4) and (6.5).

(c) (*Optional*) Using the flowchart in (a) as a guide, write a subroutine in a language that is supported on your computer that will evaluate an 8-point decimation-in-frequency FFT. Your program should accept an input file and produce an output file with the standard headers that were described in Section 1.2.

EXERCISE 6.3.5. **Two-for-One FFT**

All of the FFT algorithms that have been presented assume that the input sequence $x[n]$ is *complex*. Often, however, signals of interest are real. When this is the case, there is a further computational gain that can be realized by evaluating the DFTs of *two* real N-point sequences using *one* N-point complex DFT evaluation. Let the two real sequences be denoted by $x[n]$ and $y[n]$ and form the complex sequence

$$z[n] = x[n] + jy[n].$$

(a) Verify analytically that $X[k]$ and $Y[k]$ can be determined from $Z[k]$ using

$$X[k] = (Z[k] + Z^*[[N - k]]_N)/2$$
$$Y[k] = (Z[k] - Z^*[[N - k]]_N)/2j.$$

(b) Write a macro (called **fft241.bat**) that will evaluate the DFTs of two sequences $x[n]$ and $y[n]$ of length 16 using only one invocation of the routine **x fft**. Computing $X[k]$ and $Y[k]$ involves the use of the **x reverse, x lshift, x cshift, x conjugate, x add, x subtract**, and **x gain** functions. Verify the operation of your macro. Practically speaking, it is only necessary to produce $X[k]$ and $Y[k]$ for $k = 0, 1, \ldots, N/2$ in order to completely specify both signals. Why is this true? To simulate this feature, modify your macro (by using **x truncate** so that only $N/2 + 1$ output points of $X(k)$ and $Y(k)$ appear in the output.

(c) Develop the corresponding procedure for calculating the two inverse FFTs using one N-point inverse FFT evaluation. Your procedure should generate the two real N-point sequences $x[n]$ and $y[n]$ from $X[k]$ and $Y[k]$, which are specified for $k = 0, 1, \ldots, N/2$. Write a macro to implement this procedure and verify its operation. One approach is to first extend $X[k]$ and $Y[k]$ to their full lengths, then use these to recreate the sequence $Z[k]$, calculate the inverse transform of $Z[k]$, and finally take the real and imaginary parts of the result. The actual functions that you should use are not specified except that there should be only one invocation of **x ifft**.

EXERCISE 6.3.6. **FFT for Real Input Sequences**

The two-for-one approach of the previous exercise can be extended so that one $N/2$-point complex FFT program is used to compute the DFT of an N-point real sequence. This can be seen by looking at the flowchart shown in Fig. 6.3. It illustrates the first stage of decimation for an N-point radix-2 decimation-in-time FFT for the special case where $N = 8$. The two $N/2$-point DFTs indicated in that figure have real inputs and can be evaluated using the two-for-one FFT macro **fft241.bat** that was developed in the previous exercise.

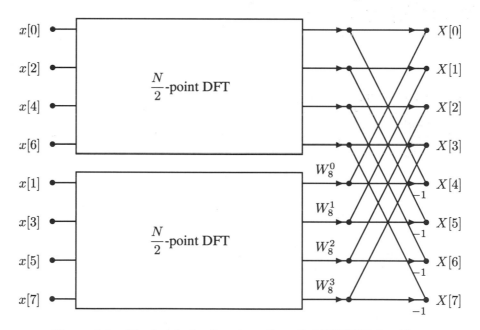

Figure 6.3.   The first decimation stage of a radix-2 DIT FFT algorithm.

(a) Use the macro **fft241.bat** as the basis for generating a new macro **fftreal.bat** that will evaluate the FFT of a 32-point real sequence. Estimate the number of multiplies and adds required using **fftreal.bat** and compare this with using a conventional 32-point radix-2 DIT FFT. What is the improvement in efficiency, if any?

(b) Write a companion macro **ifftreal.bat** that will evaluate the corresponding inverse FFT. Use your result from Exercise 6.3.5(c) as the basis for the macro.

EXERCISE 6.3.7. **Goertzel's Algorithm**

Goertzel's algorithm predates the modern rediscoveries of the FFT. Like the DFT definition it is quadratic in complexity, but it can be useful when only limited samples of the DTFT need to be evaluated or when a DFT needs to be calculated for a value of N that is prime. It is based on the observation that the DFT sample at location k

$$X[k] = \sum_{n=0}^{N-1} x[n]W_N^{nk}$$

can be expressed as one sample of the output of a linear filter

$$X[k] = y_k[N] = \sum_{m=0}^{N-1} x[m]h_k[N-m],$$

when the impulse response of the filter is chosen to be

$$h_k[m] = W_N^{-km}u[m]. \tag{6.6}$$

Because the Nth sample of the output corresponds to only one sample of the DFT, a different network has to be implemented for each value of k.

The filter in equation (6.6) has the transfer function

$$H_k(z) = \frac{1}{1 - W_N^k z^{-1}} \tag{6.7}$$

$$= \frac{1 - W_N^{-k}z^{-1}}{1 - 2\cos((2\pi k)/N)z^{-1} + z^{-2}}. \tag{6.8}$$

When the transfer function is written in the form in equation (6.7) the filter can be implemented using a first-order difference equation with a complex coefficient. It appears that one complex multiplication needs to be calculated for each value of m and k, but when the network is written in the form in equation (6.8) it can be implemented using a second-order system that has only one real feedback coefficient. The feedforward term only needs to be implemented when the Nth sample is computed. Therefore, one sample of the DFT requires $N+4$ real multiplies using the Goertzel algorithm, as opposed to $4N$ real multiplies for the direct implementation. Remember that one complex multiply is equivalent to four real multiplies and two real adds.

Create the sequence $x[n]$ given by

$$x[n] = \begin{cases} 1, & n = 0, 1, 2 \\ 0, & n = 3, 4, 5, 6 \end{cases}$$

using the *block* option in **x siggen**.

(a) Consider computing the first three samples of the 7-point DFT of $x[n]$ using Goertzel's algorithm. You will need to generate the IIR filters $H_k(z)$ for $k = 0, 1, 2$ as defined in equation (6.8). These filters can be designed using the *create file* option in **x siggen**. Use the function **x filter** to implement the filtering required by Goertzel's algorithm. Evaluate and record the values of $X(0)$, $X(1)$, and $X(2)$ using this approach. Now evaluate the DFT using **x dft**. Record and compare the values for $k = 0, 1, 2$.

(b) Compare the number of multiplies required by the Goertzel algorithm to that required by the FFT for the case where only one sample of the DFT of a 64-point sequence is desired. Describe when it is beneficial to use the Goertzel algorithm instead of the FFT.

(c) It was stated previously that multiplies and adds required by the numerator of $H_k(z)$ only needed to be performed for the Nth sample in the filtered sequence. This observation leads to additional savings in computation if all samples of the DFT are desired. The denominator polynomials for $k = 0$ and $k = N - 1$, $k = 1$ and $k = N - 2$, $k = 2$ and $k = N - 3$, and so on are identical. Hence, only about half of the recursive filters implied by the denominators need be computed. Consider this point and determine the total number of real multiplies and real adds needed to compute an N-point DFT by exploiting this property of the Goertzel algorithm.

6.4 FAST CONVOLUTION

The term *fast convolution* refers to a number of techniques for evaluating linear convolutions using discrete transforms that are implemented using fast transform algorithms. The DFT and the various FFT algorithms are often used for this purpose, but other transforms will also work. The adjective "fast" is used to contrast the efficiency between the convolution sum and the transform approaches. Evaluation of the convolution sum requires N^2 real multiplies as opposed to roughly "$4N \log_2 N + 4N$" real multiplies for the FFT transform method. For very short transforms, the convolution sum is preferable, but for large values of N, the transform techniques offer a significant reduction in the total number of operations.

EXERCISE 6.4.1. **Evaluating Linear Convolutions Using the DFT**

You have already seen that the product of two DFTs corresponds to circular convolution of their time sequences. When the length of the DFTs used is at least as long as the linear convolution of the two sequences, the circular convolution and the linear convolution are the same.

(a) Write a macro that will evaluate a fast convolution using **x zeropad,**[1] **x fft, x multiply,** and **x ifft**. It should accept as inputs two real sequences both of length $N = 2^\nu$ where ν is a positive integer. Evaluate the number of multiplies required in order to perform direct convolution and fast convolution as a function of N. Assume that each invocation of an N-point FFT requires $2N \log_2 N$ real multiplies. At what value of N is it more cost effective to use FFT-based convolution? Test your macro by convolving two block functions of length 128 and starting point zero designed using **x siggen**. To what length must the block sequences be zero padded?

(b) Assume that one sequence is the impulse response $h[n]$ of an LTI system where $h[n]$ is of finite duration with length less than 100. The input to the system is a sampled speech signal that is very long in duration. What difficulties might be encountered in implementing fast convolution for this particular case?

EXERCISE 6.4.2. **Overlap-and-Add Method**

Fast convolution can be difficult to apply when one of the sequences to be convolved is much longer than the other.

(a) Generate a triangular wave sequence $x[n]$ with 15 periods using **x siggen** with period 65, unity amplitude, and starting point zero. Use **x truncate** with ending value 969 to produce a sequence with a total length of 970 samples. Next design a 63-point highpass filter, $h[n]$, using **x fdesign**. Feel free to select an arbitrary value for the cutoff frequency and to use any one of the window options. Filter $x[n]$ by implementing the convolution directly, i.e., by using **x convolve**. How many multiplies and adds are required for this direct form implementation? Use **x view2** to display $x[n]$ and the filtered signal. Provide a sketch of the filtered signal.

(b) Now evaluate the convolution using the macro from Exercise 6.4.1 with $N = 1024$. You should modify the macro so that the DFT of the filter $h[n]$ is precomputed and used as an internal file in the macro. How many real multiplies and real adds are required assuming that $H[k]$ is precomputed? Display the results of the convolution using **x view**. Do they match those of the direct convolution?

(c) Consider breaking the input sequence into five nonoverlapping 194-point segments or blocks, $x_k[n]$, $k = 1, 2, 3, 4, 5$. Note that a block-wise partitioning of the input yields an equivalent expression for the convolution:

$$x[n] = \sum_{k=1}^{5} x_k[n]$$

$$y[n] = x[n] * h[n]$$

$$= \sum_{k=1}^{5} (x_k[n] * h[n]) = \sum_{k=1}^{5} y_k[n].$$

---

[1]Remember, to obtain a zero padded sequence of length N starting at zero, you should pad with zeros out to index $N - 1$.

Each of these smaller convolutions can be evaluated using fast convolution. Remember that the fast convolution will involve zero padding and should utilize FFTs. The resulting blocks will be longer than the original segments $x_k[n]$. Consequently, they partially overlap. For this reason the method is called *overlap and add*. Write a macro that will evaluate the five short convolutions of the segments and combine them to produce the output sequence $y[n]$. It is important to remember that each of the $x_k[n]$'s begins at a different location. Similarly, each of the $y_k[n]$'s also has a different starting point. The function **x extract** should be used to extract the blocks $x_k[n]$ and reposition them with a starting point of zero. Each block should then be efficiently convolved with $h[n]$ using the procedure developed in part (b). Some modification of the macro is needed to accommodate the different block size. After convolution is performed on each block, **x lshift** should be used to shift each block to its original starting point prior to adding up the blocks. Evaluate $y[n]$ using your macro and display it together with the filtered results from part (b).

(d) Repeat part (c), but modify the partitioning of $x[n]$ so that only 128-point FFTs are used in the fast convolution procedure.

EXERCISE 6.4.3. **Extracting Good Values from Circular Convolutions**

Generate a random 64 point sequence $x[n]$ using **x rgen** and a 16 point lowpass FIR filter, $h[n]$ with a cutoff frequency of $\pi/3$. Use the program **x fdesign** to design the filter with the *Hamming window* option.

(a) Use **x convolve** and **x cconvolve** to evaluate the following two convolutions:

- The linear convolution of $x[n]$ with $h[n]$.
- The circular convolution of $x[n]$ with $h[n]$. Use $N = 64$.

Display the two convolutions on the screen using **x view2** and sketch the results.

(b) The circular and linear convolution results are not equal, but you should be able to observe that some of the samples of the circular convolution are equal to some of the samples of the linear convolution. For which values of the index n are these two convolutions the same? (*Hint:* An easy way to determine this information is to look at the difference of the two convolutions using **x subtract**.)

(c) Generalize your observation in part (b). Assume that $h[n]$ has length N_1, $x[n]$ has length N_2, and N_2-point circular convolution is being performed where $N_1 < N_2$. In terms of N_1 and N_2, which points in $x[n] \oplus_{N_2} h[n]$ are identical to those in $x[n] * h[n]$?

EXERCISE 6.4.4. **Overlap-Save Method**

The *overlap-save method* for block convolution is the dual of the overlap-add method. It also addresses the problem of applying fast convolution when one segment is much longer than the other. It exploits the property observed in Exercise 6.4.3 that when two sequences of dissimilar lengths are circularly convolved, many of the samples of the circular convolution are equal to samples of the linear one. To develop the method, consider the convolution

$$y[n] = x[n] * h[n]$$

where the input $x[n]$ is long and the impulse response $h[n]$ is a relatively short N_1-point sequence. The approach consists of two steps:

- The input is first divided into length-N overlapping blocks. For example, the first block $x_1[n]$ might begin at $n = -25$ and end at $n = 75$, the second block $x_2[n]$ might begin at 50 and end at 150, the third $x_3[n]$ at 125 and 225, and so on.

- Within each block the N-point circular convolution

$$x_k[n] \circledast_N h[n]$$

is performed where $N_1 < N$. This results in an initial set of corrupted points (due to the circular convolution) at the beginning of each block. These corrupted points are discarded, leaving only the noncorrupted points. These input blocks should be chosen to overlap by the number of samples that are discarded so that all of the samples of the linear convolution are computed. If the block lengths N and the block overlaps are chosen correctly, the resulting uncorrupted points will form contiguous blocks that can be abutted to form the correct output sequence.

Write a macro for implementing the overlap-save convolution method. It should use the **x fft, x multiply, x ifft, x extract, x lshift**, and **x add** functions. Use your macro to convolve the sequences $x[n]$ and $h[n]$ from Exercise 6.4.2 using 256-point FFTs. How many FFTs need to be performed?

EXERCISE 6.4.5. **Use of the DFT for Deconvolution**

Deconvolution is the process of recovering one signal that has been convolved with another (often a distortion). Consider the relation

$$y[n] = x[n] * h[n].$$

The deconvolution problem may be stated as follows: given $y[n]$ and $x[n]$, find $h[n]$.

When DFTs are used for deconvolution, the quality of the results may be limited by circular convolution effects if special care is not taken. This exercise addresses this issue. Generate the two 32-point sequences

$$x[n] = (0.98)^n u[n]$$
$$y[n] = (0.98)^n u[n] + 0.5(0.98)^{n-20} u[n - 20]$$

for $n = 0, 1, 2, \ldots, 31$ using **x siggen, x gain, x lshift**, and **x add**.

(a) Compute the 32-point DFTs of $x[n]$ and $y[n]$ using **x fft** and sketch their DFT magnitudes. Perform the deconvolution by evaluating the inverse DFT of

$Y[k]/X[k]$ using **x** **divide** and **x** **ifft**. Use **x** **view** to display the real and imaginary parts of the inverse DFT. Sketch these results.

(b) Repeat part (a) using 64-point DFTs. Here it will be necessary to use **x** **ze-ropad**. Describe the effect of the DFT length on the accuracy of your result.

6.5 APPLICATIONS AND RELATIVES OF THE DFT

EXERCISE 6.5.1. **The Zero-Phase DFT**

The DFT is a transform performed on causal sequences. Since the summation that defines it begins at $n = 0$ and ends at $n = N - 1$, its direct use on zero-phase sequences is precluded. Nonetheless, the DFT can be modified to calculate N samples of the DTFT of a zero-phase (noncausal) sequence by a simple two-step process. To illustrate this point consider an odd-length symmetric sequence $x[n]$ where

$$x[n] = \begin{cases} x[-n], & n = 0, 1, \ldots, (N-1)/2 \\ 0, & \text{otherwise.} \end{cases}$$

1. Shift $x[n]$ so that its center sample is at $n = 0$. This is illustrated in Fig. 6.4a for the case $N = 7$. Next circularly extend $x[n]$ to form $\tilde{x}[n]$, as illustrated in Fig. 6.4b. The circularly shifted sequence $\tilde{x}[n]$ should again lie in the range $0 \le n \le N - 1$.

2. Take the N-point DFT of $\tilde{x}[n]$.

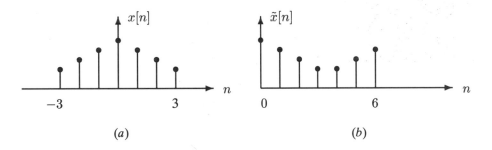

(a)   (b)

Figure 6.4.   An illustration of circular extension. (a) Zero-phase sequence.
(b) Causal sequence obtained by circular rotation ($N = 7$).

(a) Write a macro that implements the zero-phase DFT for a 15-point symmetric sequence. The functions **x** **lshift**, **x** **add** and **x** **extract** can be used to form $\tilde{x}[n]$. Display $\tilde{x}[n]$ and verify that it is correct. Next use **x** **dft** to compute the DFT. To test your program, generate 15-point lowpass and highpass filters using **x** **fdesign**. The choice of cutoff frequency and window type is left to your discretion. The output of **x** **fdesign** will be a linear-phase filter. This can be converted to a zero-phase design by shifting with **x** **lshift**. Test your macro by applying it to these sequences. Display the sequences (using **x view**) and sketch their magnitudes and phases and the real and imaginary parts. Note that plots for zero-phase signals often have phases that are nonzero. When the DTFT of a zero-phase sequence is purely real, the phase can have a value of π or $-\pi$ whenever the Fourier transform is negative. When the DTFT is purely imaginary, the phase can assume values of $\pi/2$ or $-\pi/2$.

(b) Frequency-response plots for a finite length sequence can be obtained by zero padding the sequence to some large length N and taking the FFT. These N points can then be displayed as a continuous signal to produce a representation of the DTFT. This, in essence, is how **x dtft** works. Consider a similar routine that will take the zero-phase DTFT of a symmetric or antisymmetric sequence following the procedure described in part (a). Write a macro to display the zero-phase DTFT given a 15-point symmetric or antisymmetric sequence. In this case the length of the DFT will be longer than the length of the 15-point sequence. This can be accomplished by zero padding. Use **zeropad** to pad the sequence out to a total length of 128 and take the 128-point DFT. Should the zeros be added before or after the circular rotation? Use **x dtft** to display the real and imaginary parts of your input. Next use your macro and **x look** to do the same. The **x look** function will display the output as a continuous plot. Provide sketches of all plots.

EXERCISE 6.5.2. **Phase Unwrapping**

The phase response, $\Phi(\omega)$, of a system is a multivalued function; replacing $\Phi(\omega)$ by $\Phi(\omega) + 2\pi m$ (where m is an arbitrary integer) has no effect on the Fourier transform. Certain applications require that this ambiguity in the phase function be resolved to produce a continuous (i.e., smooth) phase function. This process is called *phase unwrapping*. Phase unwrapping adds or subtracts multiples of 2π to $\Phi(\omega)$ at each value of ω to make it maximally smooth.

(a) To examine the concept of phase unwrapping, create a 4-point ramp sequence using **x siggen**. Extend it to 32 points by zero padding out to index 31 using **x zeropad** and evaluate its DFT using **x fft**. Explicitly compute files containing the magnitude and phase responses using the **x mag** and **x phase** functions. Display the phase using **x view** and sketch it. Carefully examine the phase response and identify all of the 2π discontinuities. Record these locations.

(b) Perform phase unwrapping manually by adding or subtracting appropriately shifted block functions with amplitude chosen to be multiples of 2π. This can be done using the *block* option in **x siggen** to generate the blocks, and the **x gain** and **x lshift** functions to scale and move the blocks appropriately. Sketch and display the unwrapped phase.

(c) Recombine the unwrapped phase from part (b) and the magnitude sequence obtained in part (a). This can be done by using **x gain** with the *gain* parameter 0   1 to create a purely imaginary phase sequence. This can then be added

to the purely real magnitude sequence using **x add** resulting in a complex sequence in polar form. The function **x cartesian** can be used next to convert
the sequence into its real and imaginary parts. Use **x polar** to recalculate the
magnitude and phase and display the phase using **x view**. Explain what happened to the phase of the signal.

(d) Use the *square wave* option in **x siggen** to generate one period of a square wave
with a pulse length of 3 and a starting point at zero. Compute its DTFT using **x
dtft** and display the magnitude and phase. You should observe discontinuities
of π in the phase. Unlike the previous example, these discontinuities are not due
to the fact that the phase is a multivalued function. Explain why they are present.

EXERCISE 6.5.3. The Discrete Cosine Transform (DCT)

The discrete cosine transform (DCT) is a close relative of the DFT that has found
applications in speech and image coding. One form of the DCT is a real transform
that is defined by

$$X_c[k] = e[k] \sum_{n=0}^{N-1} x[n] \cos\left(\frac{\pi(2n+1)k}{2N}\right)$$

where

$$e[k] = \begin{cases} \frac{1}{\sqrt{2}}, & k = 0 \\ 1, & k \neq 0. \end{cases}$$

The corresponding inverse DCT is similar.

$$x[n] = \frac{2}{N} \sum_{k=0}^{N-1} e[k] X_c[k] \cos\left(\frac{\pi(2n+1)k}{2N}\right).$$

(a) Express the DCT of $x[n]$ in terms of a DFT of a symmetric sequence that is related
to $x[n]$.

(b) Using this relationship, produce a macro to implement the DCT using the functions **x reverse, x lshift, x cshift, x fft,** and **x add**.

(c) Experiment with your program to determine if some of the well-known DFT
properties also hold for the DCT. For example, what is the DCT of $ax[n] + bv[n]$?
If $x[n]$ is shifted, what happens to its DCT? If two sequences are convolved, do
their DCTs multiply?

EXERCISE 6.5.4. The Discrete Hartley Transform (DHT)

The discrete Hartley transform (DHT) is another real transform and is defined as
the difference of the real and imaginary parts of the discrete Fourier transform. Thus

$$X_h[k] = \sum_{n=0}^{N-1} x[n] \left(\cos(\frac{2\pi nk}{N}) + \sin(\frac{2\pi nk}{N})\right).$$

The DHT is its own inverse (apart from a scaling factor). The convolution

$$y[n] = x[n] * h[n]$$

in the time domain becomes the following operation in the transform domain:

$$Y_h[k] = X_h[k]H_{he}[k] + X_h[[-k]]_N H_{ho}[k]$$

where $H_{he}[k]$ and $H_{ho}[k]$ are the even and odd parts of the Hartley transform of $h[n]$ and where $[[\cdot]]_N$ means to evaluate the argument modulo N.

(a) Create a macro **hartley.bat** to compute the DHT and inverse DHT. You should assume that the input sequence is real and of length 32. One of the simplest approaches uses the **x fft**, **x realpart**, **x imagpart**, and **x subtract** functions. Test your macro by taking the forward and inverse DHT of a random sequence.

(b) Describe a procedure for implementing a linear convolution using your DHT program.

(c) Write a macro that implements linear convolution using the DHT. This will use the function **hartley.bat** that you developed in part (a) along with the **x reverse**, **x add**, **x subtract**, **x gain**, **x lshift**, and **x cshift** functions.

Comment. Detailed comparisons of the relative complexity of the DFT and the DHT have shown that when optimally implemented they require exactly the same number of multiplies and nearly the same number of adds. (The DHT actually requires a few more additions.)

EXERCISE 6.5.5. **Transform Coding**

Coding is the process of limiting the number of bits required to represent a sequence. Chapter 4 presented some exercises in which quantization of sequence values was used to code a waveform. This exercise considers, as an alternative, the quantization of transform values. Let $x[n]$ denote the signal that is stored in file **sig2** and let $\hat{x}[n]$ denote its encoded representation. In general transform coding is performed in three steps:

(i) The transform of $x[n]$ is computed. There are several transforms that are popular in practice such as the DFT and DCT.

(ii) The transform values are quantized. If the transform is complex, this means that both the real and imaginary parts must be quantized.

(iii) The inverse transform of the quantized transform values is computed.

(a) As a first step, consider quantizing $x[n]$ in the conventional way to form $\hat{v}[n]$. Use **x quantize** to quantize $x[n]$ to 17 levels with minimum and maximum amplitudes of -4.82 and 4.82, respectively. Display $\hat{v}[n]$ and evaluate its SNR using **x snr**.

(b) Now consider quantizing the coefficients of a transform as opposed to the sequence values themselves. Compute the DFT of $x[n]$ using **x fft** and put the real and imaginary parts in separate files using **x realpart** and **x imagpart**. Separately quantize the real and imaginary parts using **x quantize** to 17 levels. Use amplitude ranges of -80 to 80 for the real part and -61 to 61 for the imaginary part. Next recombine the quantized signals into one complex signal. This may be done by multiplying the imaginary part of the signal by "j" and adding the sequences. Use the **x gain** function with *gain* of 0   1 and the **x add** function to perform these operations on the computer. Next, compute $\hat{x}[n]$ using **x ifft** to take the inverse DFT. Display and sketch $\hat{v}[n]$ and $\hat{x}[n]$ using **x view2**. Compute the SNR and compare it to the SNR found in part (a).

A true transform coder typically employs *adaptive bit allocation*, a procedure in which the number of levels used to represent each sample of the transform is determined adaptively based on a prescribed formula. The total number of bits used to code the sequence block is fixed initially. The formula distributes the bits to the samples based on some criterion such as energy values. In this case, assume the very simple (nonadaptive) formula in which all $X[k]$ for $k >$ 31 are discarded. Thus to measure the total number of bits used to code this sequence block, add the number of bits used to quantize $X[k]$ for k in the range $0 \leq k \leq 32$. Note that due to Hermitian symmetry, only half of the 128-point DFT is needed to uniquely represent $x[n]$. However, there are two components involved, the real and imaginary parts. Thus 32 samples of the real part and 32 samples of the imaginary part are considered when determining the total number of bits. Find the number of bits required to represent $x[n]$ using the proposed transform coding method. How many bits were required to represent $x[n]$ using the approach outlined in part (a)?

(c) Apply the transform coding method outlined in part (b) using the discrete cosine transform defined in Exercise 6.5.3 and compute the SNR. Note that the DCT produces a real transform with different properties. This calls for modification to the bit allocation formula and to the quantizer amplitude range.

(d) Apply the transform coding method using the discrete Hartley transform defined in Exercise 6.5.4 and compute the SNR. Again, the DHT produces a real transform, but has structure different from that of the DCT and DFT. Make the appropriate modifications in bit allocation and quantizer range.

EXERCISE 6.5.6. **The Chirp z-Transform (CZT)**

The chirp z-transform is an algorithm for evaluating equispaced samples of the z-transform of a finite length sequence with FFT-like efficiency. It accomplishes this by mapping a DFT evaluation into a convolution, which is then implemented using fast convolution techniques (i.e., by using the FFT).

The CZT evaluates the sequence

$$X_k = X(z)|_{z=AW^k}, \qquad k = 0, 1, \ldots, M - 1$$

$$A = |A_0|e^{j\phi_0}$$
$$W = |W_0|e^{j\theta_0}.$$

If $|W_0| \neq 1$ the samples lie on a spiral contour with an angular separation of θ_0. The complex number A controls the location of the first sample. By making use of the identity $nk = \frac{1}{2}n^2 + \frac{1}{2}k^2 - \frac{1}{2}(n-k)^2$ this sequence can be written as

$$X_k = W^{-k^2/2} \sum_{n=0}^{N-1} x[n] A^{-n} W^{-n^2/2} W^{(n-k)^2/2}.$$

This can be evaluated in three steps:

(i)   Form the sequence $g[n] = x[n] A^{-n} W^{-n^2/2}$.

(ii)   Convolve $g[n]$ with $W^{n^2/2}$ using a high-speed technique.

(iii)   Multiply the result by $W^{-k^2/2}$.

(a) Sketch the locations of the sample locations in the z-plane that would be computed by the CZT algorithm. Indicate clearly the effect of the parameters $|A_0|$, ϕ_0, $|W_0|$, θ_0, and M.

(b) Earlier in this chapter you explored techniques for evaluating linear convolutions of sequences of finite length using DFTs. The situation here is a little different. Step (ii) of the CZT procedure requires that $g[n]$, a sequence of length N, be convolved with $W^{n^2/2}$, a sequence of infinite length. However, the convolution only needs to be evaluated for M different values of its argument. This means that only a limited number of samples of $W^{n^2/2}$ are needed. Show that the convolution in step (ii) can be performed using only the samples of $W^{n^2/2}$ between limits N_L and N_U and determine the values of these parameters.

(c) Write a macro that will evaluate a 31-point CZT and verify its operation by using it to evaluate a DFT. Note that the term W^{-n^2} is equivalent to $e^{\alpha n^2 + j\theta_0 n^2}$, which can be created using the DSP functions. This can be done by generating a ramp function for n and using **x gain** and **x nlinear** to create $\alpha n^2 + j\theta_0 n^2$. The function **x nlinear** may now be used to perform the exponentiation needed to produce W^{n^2}.

(d) Compare the number of real multiplies and real adds of this program with that of a direct DFT evaluation and Goertzel's algorithm for evaluating the 31-point DFT.

(e) Use **x fdesign** to design a length-32 lowpass filter with cutoff frequency $.45\pi$. Modify the CZT macro to evaluate 100 samples of the discrete-time Fourier transform of this lowpass filter between the frequencies 0.3π and 0.6π.

6.6 REFERENCES

[1] J. W. Cooley and J. W. Tukey, "An Algorithm for the Machine Calculation of Complex Fourier Series," *Mathematics of Computation*, Vol. 19, No. 90, pp. 297–301, 1965.

[2] I. J. Good, "The Interaction Algorithm and Practical Fourier Analysis," *J. Royal Statistical Society B*, Vol. 20, pp. 361–372, 1960.

[3] M. T. Heideman, D. H. Johnson, and C. S. Burrus, "Gauss and the History of the Fast Fourier Transform," *IEEE ASSP Magazine*, pp. 14–21, Oct. 1984.

Design Projects 7

7.1 INTRODUCTION

The previous chapters review many of the basic concepts of digital signal processing. The associated exercises are there to reinforce and solidify the concepts. The projects in this chapter are quite different from the exercises in both their form and their intent. In form they are considerably more open ended; everyone who does one of the projects will probably do it somewhat differently. Their intent is to challenge the reader's creativity and force him or her to integrate many of the concepts discussed in the earlier chapters. Since a major effort may be required to complete the projects, careful thought and planning will be rewarded.

7.2 PROJECTS

Project 1. *Signal Transmission*

Many problems in communications involve sending and receiving signals without distorting the received signal. This scenario is illustrated in Fig. 7.1. The middle block in that figure represents a transmission channel with characteristics that can distort the transmitted signal. Therefore, the input, $x[n]$, is first preconditioned in the analysis block so that the signal $y[n]$ will pass through the channel with minimum distortion. The synthesis block receives $\hat{y}[n]$ and attempts to reconstruct the original input, $x[n]$.

Communications channels often suffer from nontrivial distortions involving crosstalk, amplitude and phase distortions, and echoes. In this project, the channel distortion is represented by a simple linear convolution that can be implemented using the command

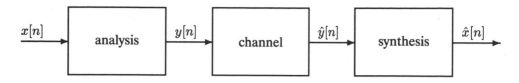

Figure 7.1. Block diagram of a communications system.

$$\mathbf{x \quad convolve \quad input \quad ccc \quad output}$$

where **ccc** is the impulse response of the channel and is included with the software. (Note, however, that the channel distortion in this project is not representative of typical channel distortions encountered in telecommunications).

The goal of this project is to design a front-end *analysis* and back-end *synthesis* system (as shown in the first and last blocks of Fig. 1) that will allow a random bandpass signal to pass through the channel and to be reconstructed as accurately as possible.

To begin, generate the input signal, $x[n]$, in the following way. Use **x rgen** to create a random sequence of length 169 with 1 as the seed value. Next convolve it with a 32-point FIR highpass filter with a cutoff frequency of 0.6π (obtained by using the *Hanning window* option in **x fdesign**). This resulting signal, $x[n]$, will have a length of 200.

Now design analysis and synthesis systems in the form of DSP macros that will allow $x[n]$ to be reconstructed at the synthesis output. The analysis system should not expand the length of the signal, i.e., $y[n]$, like the input, should be a 200-point sequence. Furthermore, your analysis and synthesis systems are not to contain any modulators. More specifically, they should not use the **x multiply** or **x cexp** function. Evaluate the performance of the system by using **x snr** to compute the signal-to-noise ratio (SNR). The challenge is to achieve the highest SNR(dB) possible.

Project 2. *Scrambling and Unscrambling*

In some government and corporate offices it is routine policy to scramble confidential messages at the transmitter to prevent eavesdroppers from acquiring privileged information. At the receiver, the signal is unscrambled so that the intended recipient of the message receives a clear undistorted signal. Signal scrambling systems are often used for voice communications where the scrambled signal is converted from digital to analog form before transmission. They find use in other applications as well. Systems of this type typically perform the scrambling digitally, convert the signal to continuous-time, resample the analog signal at the receiver, and unscramble it digitally.

Here two methods for digitally scrambling a voice signal are investigated: time-domain scrambling and frequency-domain scrambling. Time-domain scrambling consists of dividing the samples of the signal into small blocks and permuting the order of the blocks. Frequency-domain scrambling is similar except that the signal spectrum is divided into several disjoint frequency bands that are then permuted in frequency. In practice the time domain and/or frequency domain segments are permuted randomly using a pseudo-random keying sequence.

Two aspects of a scrambling system are particularly important. First, the scrambled signal should be transformed so that an eavesdropper cannot understand its message. Ideally, this means the system should attempt to produce a scrambled noise-like signal. Second, the proper recipient of the message should be able to reconstruct the original signal from the scrambled one as accurately as possible. In this project, you will design systems for time-domain and frequency-domain scrambling.

To begin, generate a 256-point periodic triangular wave with a peak amplitude of 1 and a period of 16 for values of n in the range $0 \leq n \leq 255$. Then filter this signal with an FIR lowpass filter with a cutoff frequency of $\pi/3$. Use the *Hamming window* option in **x fdesign** to construct the filter. Specify its length to be 31 and use **x lshift** to move its starting point to $n = -15$. Next, filter the triangular signal and use the **x extract** function to extract that portion of the filtered signal in the range $0 \leq n \leq 255$. This signal will be the input, $x[n]$. Display and sketch $x[n]$ and $|X(e^{j\omega})|$. What is its bandwidth (or highest frequency component)?

Frequency-Domain Scrambling.   Design a scrambling/unscrambling system based on frequency-domain scrambling. The input signal should be split into three signals—a lowpass, bandpass, and highpass signal—such that their spectra do not overlap. The lowpass, bandpass, and highpass signals should then be scrambled in frequency by modulating the filtered signals using **x cexp**. Note that bandpass filters may be designed by multiplying a lowpass filter by an appropriate cosine function.

The scrambled signal, $x_s[n]$, should have no frequency components at frequencies greater than $\pi/2$. You should carefully consider the overall problem and design the filters and modulator frequencies accordingly. For the purpose of this project it will be sufficient to establish a fixed permutation of the frequency bands as opposed to randomly scrambling them. Your objective is to determine the modulator frequencies, filters, and frequency permutation that results in the *lowest* correlation between $x[n]$ and $x_s[n]$. Restrict all filters to have unity gain and to be of length 65 in the range $-32 \leq n \leq 32$.

Write a macro that implements the scrambling system. Evaluate its performance by computing the maximum value of the cross-correlation (see Exercise 2.4.7) between $x[n]$ and $x_s[n]$. Remember the intention here is to prevent $x[n]$ from being recognized from $x_s[n]$. Your scrambling macro, however, should not expand the bandwidth of the signal beyond $\pi/2$. Display and sketch $x_s[n]$ and $|X_s(e^{j\omega})|$ and explicitly verify that your scrambling system has not increased the signal bandwidth.

Now write a macro for reconstructing the signal from $x_s[n]$. The reconstruction macro will also involve the design of filters and modulators. In all cases use filters of length 65. To evaluate the performance of the overall system, use **x snr** to compute the signal-to-noise ratio between the original signal and the reconstituted one. Your objective should be to design the unscrambler so that the overall system has the highest SNR possible.

Time-Domain Scrambling.   In this part of the project, you are to write a macro for a scrambler that permutes blocks (of length 51) in the time domain. Explicitly scramble these blocks within your macro using the **x extract** function and attempt to find the permutation that results in the lowest correlation between $x[n]$ and $x_s[n]$.

Now assume that the scrambled signal $x_s[n]$ is passed through a bandlimited channel that you can simulate by filtering $x_s[n]$ with a 65-point FIR lowpass filter with a cutoff frequency of $2\pi/3$ and starting at $n = -32$. Call this filtered signal $x_f[n]$ and display and sketch both $|X_s(e^{j\omega})|$ and $|X_f(e^{j\omega})|$. Observe that time-domain scrambling has an apparent disadvantage in that it increases the bandwidth of the signal. However, it is possible to avoid this problem and produce a time-domain scrambler that will work over bandlimited analog channels. Since $x[n]$ is bandlimited to $\omega_c = \pi/3$, resampling (or downsampling) it at its Nyquist rate results in a new signal that occupies the full spectrum, $(-\pi \le \omega \le \pi)$. If scrambling and unscrambling are performed on this decimated signal, no increase in bandwidth occurs. This is true because the signal already has the maximum bandwidth possible (i.e., its spectrum extends from 0 to π.) Reconstruction can be achieved by upsampling and interpolating to the original rate. Based on this observation, draw a block diagram of a new time-domain scrambling system. The scrambling system output should be at the same rate and have the same bandwidth as the input. Also draw the block diagram for the corresponding unscrambler.

Using the block diagrams to guide you, write macros that implement the time-domain scrambler and unscrambler. Record the overall SNR and the cross-correlation between the scrambled and unscrambled signals as measures of how well your system works.

Project 3. *Quantizer Design*

Quantization is the process in which sample values of an arbitrary signal, $x[n]$, are replaced by discrete amplitude levels $\hat{x}[n] = Q\{x[n]\}$. This topic was discussed in Chapter 4 in the context of D/A and A/D conversion. You may wish to refer to Section 4.2 for a review of this material.

When the accuracy of the signal representation is limited, its capacity for holding information is also limited. For this reason quantization is very important in digital processing. To model the effect of quantization it is convenient to model $x[n]$ as a random variable characterized by its probability density function or pdf. In practice, it is often satisfactory to represent the pdf by its mean, μ, and variance, σ^2.

The objective of this project is to quantize $x[n]$ to 4 bits or 16 levels (using the **x quantize** function) in a way that produces the highest signal-to-noise ratio (SNR) possible. The different parts of this project should indicate ways to obtain improved SNR performance.

First we need to generate the test signal that will be used throughout the project. Following the procedure described in Exercise 2.4.9(a), create a zero-mean 256-point Gaussian random sequence, $x[n]$, with standard deviation $\sigma = 1$. Your sequence should extend over the interval $0 \le n \le 255$.

The simplest way to use **x quantize** to produce a 4-bit signal representation is to use the uniform quantizer defined by the minimum and maximum values of $x[n]$, X_{max} and X_{min}. This approach is used in Exercise 4.2.1. As an initial lower bound on performance, quantize $x[n]$ in this manner and compute the resulting signal-to-noise ratio using **x snr**.

Plot the histogram of $x[n]$ (as you did in Exercise 2.4.8(b)), normalize the histogram to have an area of one, and use the plot as an estimate of the signal pdf. Observe that the signal is zero mean and that its pdf goes to zero for large values of $|x|$. This suggests that tightening the minimum and maximum quantization limits could improve the SNR. The pdf indicates that samples that fall in the saturation range of the quantizer occur infrequently. By increasing the amount of overload distortion (i.e., reducing the uniform quantization range $X_{max} - X_{min}$) the step size, Δ, can be reduced and the quantization error can made smaller. Systematically vary X_{max} and X_{min} in **x quantize** and determine the values that produce the best SNR. Record these limits and the corresponding SNR value.

A closer examination of the pdf shows that small samples occur more often than large ones. It is therefore appropriate to quantize those low-amplitude samples that occur with higher probability more accurately. One way to do this is with a nonuniform quantizer. Nonuniform quantization can be implemented using a uniform quantizer in the following way. It is based on the fact that a uniform quantizer is optimal when the pdf is uniform. Therefore, if we can reversibly transform a sequence $x[n]$ with a nonuniform pdf into a new sequence $y[n]$ with a uniform pdf, $y[n]$ can be optimally quantized (resulting in $\hat{y}[n]$) using a uniform quantizer. The quantization of $x[n]$ can then be performed by inverse transforming $\hat{y}[n]$ to produce $\hat{x}[n]$. This general approach is called *companding* because of the *com*pression and ex*pan*sion operations. The whole process is illustrated in Fig. 7.2. The transformation or compression operation, which we denote as $D\{\cdot\}$, is often called the *companding law.* The expansion operation $D^{-1}\{\cdot\}$ is its inverse. Companding is a conceptually simple method of implementing a nonuniform quantizer. In practice, however, nonuniform quantizers are usually not implemented this way.

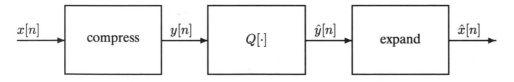

Figure 7.2. Block diagram of a companding method for implementing a nonuniform quantizer.

To improve the performance of the quantizer, consider the particular companding law defined for a zero-mean signal with $|X_{max}| = |X_{min}|$:

$$y[n] = D\{x[n]\} = X_{max} \frac{\ln(1 + \mu|x[n]|/X_{max})}{\ln(1 + \mu)} \text{sgn}(x[n]) \tag{7.1}$$

where $\mu \geq 0$ and $\text{sgn}(\cdot)$ is $+1$ when the argument is positive and -1 when the argument is negative. This companding law is called μ-law quantization. When μ is zero, the characteristic is linear, i.e., $y[n] = x[n]$. When μ is nonzero (perhaps 50, 100, 200, etc.) the resulting nonuniform quantization contains a higher density of levels near zero.

Write a macro to implement μ-law compression $(D\{x[n]\})$ and another to implement the expansion, $D^{-1}\{x[n]\}$, with μ as a free parameter that you can specify. Note that **x quantize** can be used to implement the $\text{sgn}(x[n])$ function. The **x log, x add, x gain, x divide, x multiply,** and **x siggen** functions can be used to realize the remainder of the expression. (You may wish to refer to Exercise 2.4.5(a) which also called for implementing a nonlinear equation using a macro.) Implement your companding quantizer and systematically determine the value of μ that results in the highest SNR.

The μ-law compressor does not explicitly exploit the signal statistics. Further improvement should be possible with a quantizer that directly exploits the pdf. You are challenged to design a compander that will achieve the very highest SNR possible. To give yourself maximum flexibility, implement the compander in the form of DSP-compatible *compress* and *expand* programs. Record the highest SNR value you are able to achieve.

Project 4. *Signal Extrapolation*

This project is concerned with reconstructing a long bandlimited signal when only a short segment of that signal is known. This general problem is known as *extrapolation*. Assume that you are given $y[n]$, defined as

$$y[n] = \begin{cases} c[n] & \text{if } -L \leq n \leq L \\ 0 & \text{otherwise,} \end{cases} \tag{7.2}$$

where $c[n]$ is bandlimited. The objective of this project is to recover the whole signal $c[n]$ from the time-limited segment $y[n]$.

At first glance this may seem impossible. The fact that $c[n]$ is bandlimited to some frequency $\omega_0 < \pi$, however, permits the method of successive energy reduction [1, 2] to be applied. The method is based on the following observations:

1. Since $c[n]$ is bandlimited to some frequency, ω_0, $C(e^{j\omega}) = 0$ for $\omega_0 < |\omega| < \pi$. Moreover, the bandlimited condition on $c[n]$ implies that the signal must also be of infinite duration.

2. Since $y[n]$ is time limited, $Y(e^{j\omega})$ must span the full frequency range $-\pi \leq \omega \leq \pi$. If $y[n]$ is to be fully extrapolated to equal $c[n]$, then those samples that are added to lengthen $y[n]$ must also contribute to canceling those frequency components of $Y(e^{j\omega})$ that lie above $|\omega| = \omega_0$.

This suggests an iterative strategy in which we start with an initial estimate $y_0[n] = y[n]$ and improve our estimate after each iteration. The algorithm is formulated so that the ℓth estimate $y_\ell[n]$ is always a better approximation to $c[n]$ than the previous estimate

$y_{\ell-1}[n]$. The strategy is based on reducing the error energy after each iteration by forcing $Y_\ell(e^{j\omega})$ to be equal to zero in the frequency interval $\omega_0 < |\omega| < \pi$ via lowpass filtering. In addition, all estimates of the iterates $y_\ell[n]$ are set equal to $y[n]$ in the interval $-L \leq n \leq L$.

To illustrate the approach, examine Fig. 7.3, which depicts a single iteration of the algorithm as a two-step process. At the beginning of the iteration we have the estimate, $y_\ell[n]$, which we can examine in terms of the discrete-time Fourier transform, $Y_\ell(e^{j\omega})$. Unless the iteration has converged, this signal will not be bandlimited. Therefore, we bandlimit it to the frequency ω_0 by lowpass filtering. This results in the signal $V(e^{j\omega})$ as shown in the figure. If the algorithm has converged, the inverse discrete-time Fourier transform (DTFT) of $V(e^{j\omega})$, $v[n]$, will be identical to $y[n]$ in the observation interval $-L \leq n \leq L$. The second step, therefore, is to replace samples of $v[n]$ with samples of $y[n]$ in the observation interval. The result of this step is $y_{\ell+1}[n]$. By repeating these steps, $y_\ell[n]$ will converge to the minimum energy extrapolation solution, which is unique.

In this project, we will use as the ideal bandlimited signal

$$c[n] = \cos \frac{\pi}{64} n.$$

Construct a time-limited signal, $y[n]$, which contains 17 samples of this sinusoid in the range $-8 \leq n \leq 8$. This will serve as the observation signal to be extrapolated.

Using Parseval's relation, show that each iteration in the proposed algorithm reduces the error energy, e_ℓ, defined as

$$e_\ell = \sum_{n=-\rho}^{\rho} (c[n] - y_\ell[n])^2,$$

for some large, but finite value of ρ.

There are many practical issues to consider in implementing this algorithm. These include determining the length and quality of the lowpass filter, whether to implement filtering in the time domain or frequency domain, how to handle the increasing length of $x_\ell[n]$ after each iteration to avoid excessive truncation errors, how many iterations to be performed, etc. Write a macro, or a program if you wish, that will extrapolate $y[n]$ to 33 points so that $y_N[n] \approx c[n]$ for $-16 \leq n \leq 16$. If you decide to implement your system in the form of a macro, consult Section 1.3, which describes a simple macro structure for implementing an iteration. To judge the accuracy of your extrapolation, use **x snr** to compute the signal-to-noise ratio. Extrapolation methods of this type are typically very numerically sensitive and accurate extrapolations are difficult to achieve. Your objective is to design your procedure to achieve the highest SNR possible.

Project 5. *Using the Complex Cepstrum*

Systems are often designed to have a certain frequency response with little regard to the locations of their poles and zeros. Implementation considerations, however, often

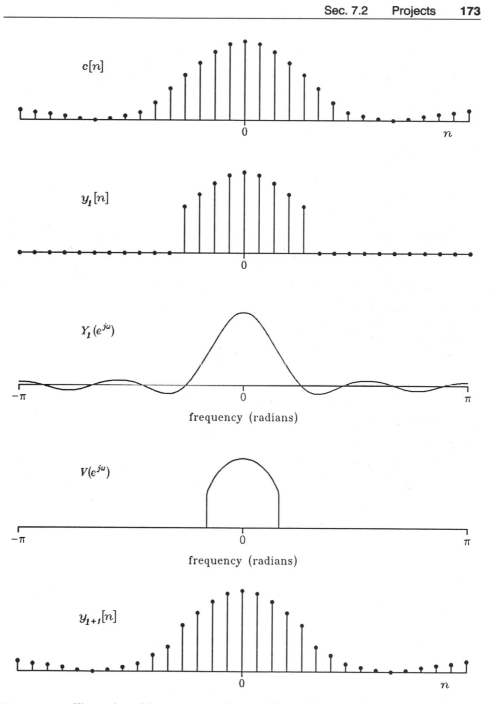

Figure 7.3.   Illustration of the two-step process used in each iteration of the successive energy reduction method.

dictate that the poles and zeros of a system must lie within the unit circle. A causal system must have its poles inside the unit circle if it is to be stable, and it must have its zeros inside the unit circle if it is to be invertible with a stable, causal inverse. For any system, however, we can find a related system that has all of its poles and zeros inside (or possibly on) the unit circle, and which has the same magnitude response as the original. When all of the poles and zeros lie inside the unit circle, the system is said to be *minimum phase*.[1]

We will first consider systems with no poles or zeros on the unit circle. In such cases the minimum-phase system can be found by replacing all of the poles and zeros of the system function that lie outside the unit circle with poles and zeros at their reciprocal locations. This procedure is known as *pole (or zero) flipping*. For example, if $H(z)$ has a pole at $z = 2$, then flipping the pole to its reciprocal location would move it to $z = 1/2$. A straightforward approach that can be used to find the minimum-phase system is to find the roots of the numerator and denominator polynomials of $H(z)$, compute the reciprocals for the outside-the-circle roots, and then reassemble the numerator and denominator polynomials using the modified roots. This procedure will break down if the polynomials are of very high order, since it may not be possible to compute their roots with sufficient accuracy.

In this project, an alternative approach is considered that makes use of the *complex cepstrum*, which is defined as follows. Suppose $h[n]$ and $H(z)$ are the impulse response and rational system function, respectively, of a linear time invariant (LTI) system. The complex cepstrum of $h[n]$ (which we will call $h_c[n]$) is the inverse z-transform of $\ln H(z)$. It has the property that roots (either poles or zeros of $H(z)$) inside the unit circle contribute to $h_c[n]$ for $n > 0$, and roots outside the unit circle contribute to $h_c[n]$ for $n < 0$. To see this, consider the simple three-point FIR filter

$$h[n] = -3\delta[n+1] + 7\delta[n] - 2\delta[n-1],$$

which has the z-transform $H(z)$. Expressing $H(z)$ in terms of its roots results in

$$H(z) = 6\left(1 - \frac{1}{3}z^{-1}\right)\left(1 - \frac{1}{2}z\right).$$

Taking the logarithm of the transform yields

$$H_c(z) = \ln H(z) = \ln 6 + \ln\left(1 - \frac{1}{3}z^{-1}\right) + \ln\left(1 - \frac{1}{2}z\right).$$

The inverse transform of $H_c(z)$, which is the complex cepstrum, can be obtained by making use of the power series expansion

$$\ln(1 - \alpha) = -\sum_{n=1}^{\infty} \frac{\alpha^n}{n}, \qquad |\alpha| < 1.$$

---

[1]The term, minimum phase, is also used to refer to systems with roots inside as well as on the unit circle.

This gives

$$H_c(z) = \ln 6 - \sum_{n=1}^{\infty} \frac{1}{n} \left(\frac{1}{3}\right)^n z^{-n} - \sum_{n=1}^{\infty} \frac{1}{n} \left(\frac{1}{2}\right)^n z^n,$$

which has the same form as the z-transform definition. Replacing n by $-n$ in the second summation allows us to recognize immediately the inverse z-transform, $h_c[n]$:

$$h_c[n] = (\ln 6)\delta[n] - \frac{1}{n}\left(\frac{1}{3}\right)^n u[n-1] + \frac{1}{n}(2)^n u[-n-1].$$

This simple example illustrates how the zeros of a finite length signal contribute to the complex cepstrum; roots outside the unit circle transform to sequences in the negative time region of the complex cepstrum and roots inside contribute to the positive region. You should generalize this example and show how the poles of a system function contribute to the complex cepstrum. This procedure can be applied to any signal, $x[n]$, not only to a system impulse response.

The complex cepstrum of a minimum-phase signal is causal and it has the same even part as the cepstrum of every other signal that shares its magnitude response. Therefore, by time reversing the negative-time portion of the complex cepstrum of an arbitrary signal and adding it to the positive-time portion, we have the complex cepstrum of the corresponding minimum-phase signal. By inverting this complex cepstrum (i.e., by taking its transform, exponentiating, and taking the inverse transform), we have the minimum-phase system itself. This is the basis of the algorithm that is the topic of this project.

(a) Write a macro that will convert an odd length, linear phase signal, $x_L[n]$, to a minimum-phase signal, $x_M[n]$ (with all its exterior roots flipped inside the unit circle), using the complex cepstrum. Design the macro to handle odd length signals as long as 201 samples. Test your procedure using the linear phase signal produced by the following procedure:

   (i)   Generate the 16-point damped exponential $v[n] = (0.85)^n$ where $0 \leq n \leq 15$ using **x siggen**.

   (ii) Let $x_L[n]$ be the 31-point sequence given by $x_L[n] = v[n] + v[-n]$. Compute $x_L[n]$ using the **x reverse** and **x add** functions.

You should reformulate the computation of the complex cepstrum in terms of the DFT so that FFTs can be used for its computation. There are many issues that have to be addressed in designing the system, including the selection of the length of the FFTs, modifications to the FFT procedure to permit it to be used with signals that do not begin at $n = 0$, etc.

Comment. Note that the test signal given is even and zero-phase. Thus its Fourier transform is real. In addition, it has no zeros on the unit circle. For now you can build these assumptions into your macro, but they will be removed in subsequent parts of the project.

(b) If your macro is working properly, the magnitudes of the Fourier transforms of $x_L[n]$ and $x_M[n]$ will be the same. Use **x polezero** to determine the zeros and Fourier transform magnitudes of $x_L[n]$ and $x_M[n]$ and verify the correct operation of your macro. Sketch the pole/zero and spectral magnitude plots for each. What steps could be taken to further increase the accuracy of the macro?

(c) Consider a 31-point input signal with zeros on the unit circle. What potential problems does this present to an algorithm based on the complex cepstrum? Generate a 31-point signal, $y_L[n]$, with several zeros on the unit circle (such as the impulse response of a lowpass filter). Execute your macro to obtain $y_M[n]$ and discuss its performance. Sketch pole/zero plots for $y_L[n]$ and $y_M[n]$.

(d) For a comparison, convert $x_L[n]$ into $x_M[n]$ by explicitly computing the roots, taking the reciprocals of the roots that lie outside the unit circle, and reconstructing the polynomial. This can be done using **x rooter, x polar, x realpart, x imagpart, x mag, x divide**, and **x rootmult.** Record your result and compare it with that in part (a).

(e) The test signal considered earlier was relatively short. Design a signal, $w_L[n]$, as described in part (a), but with a length of 199 instead of 31. Transform it to a minimum-phase signal using your macro from part (a) and also by following the procedure outlined in part (d). While the procedure that finds roots of the z-transform of the signal will inherently have difficulty for long sequences, the macro based on the cepstrum can often do the job. Modify your macro, if necessary, so that it can handle this long input signal. Display and sketch the minimum-phase signal, $w_M[n]$. Check its accuracy by comparing $|W_M(e^{j\omega})|$ with $|W_L(e^{j\omega})|$.

(f) It is often necessary to convert a nonzero phase sequence into a minimum-phase one. Procedures based on polynomial rooting will certainly work providing that the polynomial can be accurately rooted. In this part, you are to extend the complex-cepstrum-based algorithm to handle nonzero phase signals. In this case, you will no longer be able to ignore the phase and will be forced to treat it explicitly. Since the phase is a multivalued function, it should be unwrapped prior to computing the inverse Fourier transform. Phase unwrapping, as you may recall, was discussed in Exercise 6.5.2.

Write a program to perform phase unwrapping and integrate it into the cepstrum macro. To test the effectiveness of the macro, create a 24-point chirp signal, $z[n]$, using **x siggen** with *amplitude* equal to 1, *alpha1* equal to 0.001, *alpha2* equal to 0.01, *phi* equal to 0.05, and starting point $n = 0$. Display and sketch the pole/zero plot and magnitude response. Convert $z[n]$ to the minimum-phase signal, $z_M[n]$, using your macro and sketch the pole/zero plot for $z_M[n]$. Now examine the performance of the macro for long signals. Refine your macro and phase unwrapping program to accommodate long signals. Performance problems become more severe when longer signals are involved. Try to make the modified macro capable of effectively handling these longer signals. Determine the approximate signal length beyond which your macro's effectiveness noticeably deteriorates. Since $|Z(e^{j\omega})|$ should equal $|Z_M(e^{j\omega})|$, this can be used as a test for successful operation. The accuracy of your macro will also depend on the closeness of roots to the unit circle. Discuss the problems associated with this issue.

Project 6. *Nonuniform Sampling*

The sampling theorem, which was discussed in Chapter 4, stated that a bandlimited signal, $x_a(t)$, that has been sampled to form the sequence $x[n] = x_a(nT)$ can be reconstructed exactly from those samples provided that $T < \pi/\Omega_0$, where Ω_0 is the bandwidth of the signal. The minimum sampling rate for exact reconstruction, $f_s = 1/T = \Omega_0/\pi$ is called the *Nyquist rate*. The reconstruction is governed by the equation

$$x_a(t) = \sum_{n=-\infty}^{\infty} x[n]\phi(t - nT) \tag{7.3}$$

where $\phi(t)$ is an interpolation function. For ideal sampling, the interpolation is defined by

$$\phi(t) = \frac{\sin(\pi/T)t}{(\pi/T)t}. \tag{7.4}$$

Since equation (7.3) involves an infinite summation and is not realizable in practice, other (nonideal) interpolation functions are usually used and the resulting reconstruction is not exact.

The sampling theorem can be generalized to include nonuniform sampling [3]. The generalized sampling theorem states that $x_a(t)$ can be recovered from nonuniformly spaced samples provided that the average sampling rate is above the Nyquist rate. Here, we will investigate a particular form of nonuniform sampling and attempt to reconstruct the signal with high fidelity. However, since the computer environment constrains us to operate in the discrete-time domain, we will work with a densely sampled signal, $x_a(nT)$, sample it at a reduced rate, and then attempt to reconstruct it.

Using **x siggen** and **x add**, create the 128-point signal

$$x[n] = x_a(nT) = \sin\frac{\pi nT}{64} + \sin\frac{\pi nT}{32} + \sin\frac{\pi nT}{16}$$

where $T = 1$ and $0 \le n \le 127$. Using **x view** and **x dtft**, display and sketch $x[n]$ and its discrete-time Fourier transform, $X(e^{j\omega})$. Since a finite length signal cannot be truly bandlimited, assume that $x[n]$ is periodically replicated every 128 samples and thus extends infinitely in both directions. In this way $x[n]$ can properly be viewed as an oversampled periodic signal and, therefore, can be represented with fewer than 128 samples per period.

Reconstruction Based on Uniform Sampling. The signal $x[n]$ is based on $T = 1$. However, since the signal is oversampled, the sampling period T can be increased to 8 without information loss. Using **x dnsample** produce the sequence $y[n] = x_a(nM) = x[nM]$ where $M = 8$.

Write a program, called **urecon**, that will accept $y[n]$ as its input and produce an output, $\hat{x}[n]$, such that $\hat{x}[n] \approx x[n]$ based on the reconstruction equations (7.3) and (7.4). You may wish to exploit the fact that $x[n]$ is periodically replicated. This will provide a way for you to improve the accuracy of your interpolation by enabling you to effectively extend the summation indefinitely. With some careful thought, it is possible to modify equations (7.3) and (7.4) to obtain an exact reconstruction expression for a periodic signal. To evaluate the success of your program, compute the signal-to-noise ratio using **x snr**. Design your program to produce the highest SNR(dB) possible and record the best SNR that you can obtain in your write-up.

Now consider reconstructing $x[n]$ from $y[n]$ using linear interpolation. Linear interpolation simply involves connecting each sample value of $y[n]$ with straight lines. For example, in the case where $M = 2$, reconstruction via linear interpolation simply involves computing the average value of each pair of samples and inserting it between them. Determine how to implement linear interpolation (for the 1-to-8 interpolation case) using the DSP software and write a macro that performs this operation. Again design the system to produce the highest SNR possible given the linear interpolation constraint. Compute and record the SNR(dB) value for this case. Display $\hat{x}[n]$ and compare it with the results obtained previously.

Reconstruction Based on Nonuniform Sampling.   In this part, you will consider a new input signal. Using **x siggen**, create a 128-point chirp signal with *alpha1* = 0.006, *alpha2* = 0.001, and starting point at $n = 0$. As before we will call this signal $x[n] = x_a(nT)$ where T is assumed to be equal to 1. Display and sketch both $x[n]$ and $X(e^{j\omega})$.

When a signal $x_a(t)$ has an exploitable time structure, sampling can be performed at rates lower than the Nyquist rate by using nonuniform sampling. In a method proposed by Clark *et al.* [4], the time variable, t, is mapped by a function, $\vartheta(t)$, to produce a new signal, $v_a(t) = x_a(\vartheta(t))$. This new signal, $v_a(t)$, is then uniformly sampled in the conventional way to produce $v_a(n\hat{T}) = x_a(\vartheta(n\hat{T}))$. Thus, not only does this approach succeed in achieving nonuniform sampling, but it transforms the nonuniform sampling problem into a conventional uniform sampling problem. The sampling theorem now imposes the constraint that $v_a(t)$ must be uniformly sampled above the Nyquist rate associated with $v_a(t)$. In addition, the warping function $\vartheta(t)$ must be invertible so that $x_a(t)$ can be recovered from $v_a(t)$ using the relationship

$$x_a(t) = v_a(\vartheta^{-1}(t)) \tag{7.5}$$

after $v_a(t)$ has been recovered from its uniform samples.

Determine a warping function, $\vartheta(t)$, suitable for the chirp signal in this project. Based on the discussion above, write a program, **warp**, that will transform $x[n] = x_a(nT)$ into $v[n]$. Then write a program, **unwarp**, that will reconstruct $x_a(nT)$ from $v[n]$. The challenge is to design the warping program to produce the signal $v[n]$ with a small bandwidth. The warping and unwarping programs should not introduce significant distortion. Compute the signal-to-noise ratio to verify that the programs are operating properly. Record the SNR for the warp/unwarp programs and the bandwidth of $v[n]$.

After demonstrating the successful operation of the warp and unwarp programs, downsample $v[n]$ by an appropriate factor and use a modified version of **urecon** to reconstruct it. This entire system represents nonuniform sampling. Evaluate and record the SNR for the overall system. How does the number of samples required by this method compare with the number required if $x_a(t)$ is uniformly sampled at its Nyquist rate.

Project 7. *30-Point FFT Programs*

Chapter 6 reviewed a number of algorithms for evaluating the discrete Fourier transform. In this project you will write two programs in a high-level language supported in your computing environment that will evaluate a 30-point DFT using the decimation-in-time Cooley–Tukey algorithm and the prime factor algorithm with decimations of 2, 3, and 5. These programs do not need to be capable of evaluating DFTs of any other lengths.

Cooley–Tukey Algorithm

(a) Draw a detailed flowchart of the algorithm. The procedure consists of three stages.

  (i) First fifteen 2-point DFTs need to be evaluated on pairs of input samples.

  (ii) These outputs must be multiplied by the appropriate twiddle factors, and ten 3-point DFTs of the products must be evaluated.

  (iii) Finally each of the outputs of the 3-point DFTs must be multiplied by additional twiddle factors, and then six 5-point DFTs should be evaluated.

Your flowchart should clearly indicate which signal values are input to the various short DFTs. The appropriate coefficients should also be indicated.

(b) Now program your algorithm. Since this program does not need to handle arbitrary length sequences, you should feel free to use "straight-line code." For example, you do not need to program general-purpose bit reversers and data scramblers. Instead your program might look something like

$$\vdots$$
$$g(3) \;=\; f(8) + \text{coef}(3) * f(29)$$
$$g(4) \;=\; f(9) - \text{coef}(4) * f(28)$$
$$\vdots$$

(c) Verify the operation of your program.

Prime Factor Algorithm. This part of the project involves writing a program for evaluating a 30-point prime factor FFT, similar to the one in the previous section, so that the efficiencies of the two FFTs can be compared. For the prime factor algorithm, which was described in Chapter 6, you should use the parameters:

$$\mu_1 = 15; \qquad \mu_2 = 10; \qquad \mu_3 = 6$$
$$R_1 = 1; \qquad R_2 = 1; \qquad R_3 = 1.$$

As with the Cooley–Tukey FFT, the algorithm will involve a number of 5-point, 3-point, and 2-point DFTs, but the connections between them are different.

(d) Draw a complete flowchart of the procedure, clearly indicating the inputs to each DFT.

(e) Code your procedure in the same high-level language that you used in part (b).

(f) Verify the operation of your program.

(g) How do the 30-point Cooley–Tukey and 30-point prime factor programs compare with respect to execution time, program length, and programming difficulty?

Project 8. *Split-Radix FFT*

The split-radix FFT is the most efficient of the Cooley–Tukey-type FFT algorithms. It was discovered by Duhamel and Hollmann in 1984 [5].

The discrete Fourier transform of the sequence $x[n]$ is defined as

$$X[k] = \sum_{n=0}^{N-1} x[n] W_N^{nk} \qquad k = 0, 1, \ldots, N - 1.$$

If N is a power of 2, the DFT can be broken into two sums—one over the even-indexed samples and one over the odd-indexed samples. Rearranging them slightly as

$$X[k] = \sum_{n=0}^{\frac{N}{2}-1} x[2n] W_{N/2}^{nk} + W_N^k \sum_{n=0}^{\frac{N}{2}-1} x[2n+1] W_{N/2}^{nk}$$

allows each of the sums to be recognized as an $N/2$-point DFT. Each of them can be broken up again into two $N/4$-point DFTs, and by continuing this recursive factorization until only length-2 DFTs remain, the standard decimation-in-time FFT algorithm is derived.

This approach is not limited to factors of 2. If N is a power of 4, each step of the algorithm can factor the transform into four quarter-length DFTs. This leads to the radix-4 decimation-in-time FFT algorithm. The split-radix algorithm is a hybrid between the two. It factors the even-indexed elements as a radix-2 FFT, while the odd-indexed elements are factored as a radix-4 FFT. Thus, the DFT can be expressed as:

$$X[k] = \sum_{n=0}^{N/2-1} x[2n]W_{N/2}^{nk} + W_N^k \sum_{n=0}^{N/4-1} x[4n+1]W_{N/4}^{nk}$$

$$+ W_N^{3k} \sum_{n=0}^{N/4-1} x[4n+3]W_{N/4}^{nk}, \qquad k = 0, 1, 2, \dots, N-1.$$

Because the first sum is periodic in k with period $N/2$ and the other two with period $N/4$, we can write

$$X[k] = G[k] + W_N^k H[k] + W_N^{3k} I[k]$$

$$X\left[k + \frac{N}{4}\right] = G\left[k + \frac{N}{4}\right] - jW_N^k H[k] + jW_N^{3k} I[k]$$

$$X\left[k + \frac{N}{2}\right] = G[k] - W_N^k H[k] - W_N^{3k} I[k]$$

$$X\left[k + \frac{3N}{4}\right] = G\left[k + \frac{N}{4}\right] + jW_N^k H[k] - jW_N^{3k} I[k]$$

The first stage of decimation is shown in Fig. 7.4 (for an 8-point transform). The approach is continued until only 2-point DFTs remain. The 2-point DFTs are identical to those in the standard Cooley–Tukey algorithm.

(a) Draw the complete flowchart of a 16-point split-radix FFT.

(b) Program a split-radix FFT that will work for a general $N = 2^M$ using a high-level language that is supported in your computing environment.

(c) How many multiplications and additions are required for a length $N = 2^M$ split-radix FFT? Prepare your answer in the form of a table with entries for lengths $16, 32, 64, \dots, 1024$. How do these numbers compare with the numbers required for a normal radix-2 algorithm? With a radix-4 algorithm?

(d) Run both the split-radix algorithm and a normal radix-2 FFT on sequences of length $16, 32, 64, \dots, 1024$ and compare their computation times.

Project 9. *FFT Spectrum Analyzer*

The purpose of this project[2] is to design a spectrum analyzer that computes short-time spectra for time-varying signals, such as speech, over a very limited range of frequencies. Assume that the input signal is real-valued and is sampled at 8000 samples per second. We want to extract various segments of the signal and analyze them for frequency components in a 100-Hz range above some "analysis frequency," f_a. We will assume that the sampling is "ideal"—no aliasing, perfect A/D conversion, etc.

---

[2]This project was originally written by Professor James McClellan.

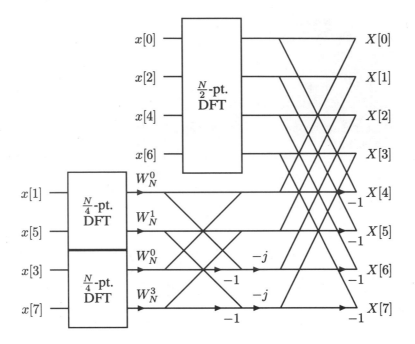

Figure 7.4.   The first stage of decimation for a decimation-in-time split-radix FFT ($N = 8$).

Part 1.   Since the signal $x[n]$ is time-varying, consider the following scheme. The signal is multiplied by an N-point Hamming window that ends at sample m to produce the signal

$$x_m[n] = x[n]w[m - n]$$

whose spectrum is then computed using an N-point DFT. The resulting spectrum will be a function of both the window position m and the frequency index k. This is sometimes called a *sliding window DFT*. The DFT samples can be identified with samples of the Fourier transform of the original continuous-time signal.

  (a) If we want samples of the spectrum to be at a spacing of less than 5 Hz, what is the minimum length radix-2 FFT required to do this directly? How many outputs of the FFT will be used to cover the band of interest?

  (b) Now consider the sliding window DFT. If we track one of the DFT outputs, say the kth as a function of the window position m, we see that it is the output of a narrow (complex) bandpass filter whose bandwidth is on the order of a few hertz. It is therefore possible to downsample the output of the sliding window DFT by a large factor. (This only removes the "carrier" associated with the center frequency of the channel.) Determine this downsampling factor assuming that aliased components will be rejected by the sidelobes of the Hamming window (approximately -40 dB) prior to aliasing. Derive a formula for the frequency response of the effective bandpass filter formed by tracking the kth DFT sample.

(c) If the length of the DFT is N, then each calculation performed by the sliding window DFT involves a block of N signal points. If there is no downsampling, then successive blocks are offset by one sample. What is the offset between blocks when the output is downsampled?

Part 2. One problem with the direct FFT approach is that only a small fraction of the outputs, $X_m[k]$, are of interest, so we are obviously wasting a lot of computation when the FFT calculates all the frequency-domain outputs. An alternate approach is to use a complex bandpass filter (BPF) to filter and downsample the input signal prior to the FFT analysis. One way to implement this kind of spectrum analysis system is to filter (BPF) out the band of interest, multiply it by an exponential to shift it to baseband, downsample it, compute an FFT, and downsample again. But there is an alternative that avoids the multiplication by the complex exponential: For certain values of f_a the frequency shift to baseband can be done directly by the downsampling process.

(d) Derive the condition on f_a so that the frequency shift to baseband can be accomplished by downsampling. Verify that the value $f_a = 1$ kHz satisfies this condition. Does your result assume an ideal BPF? How does your result depend on the transition width of the BPF?

(e) Using the idea contained in the preface to Part 2 above, design a complete system to perform the limited range frequency analysis. Use the value $f_a = 1$ kHz. The system should be the cascade of four blocks, as illustrated in Fig. 7.5: BPF, downsampler, sliding-window FFT, downsampler. (Actually, the sliding-window FFT does its decimation by skipping over a large fraction of the FFT block length.)

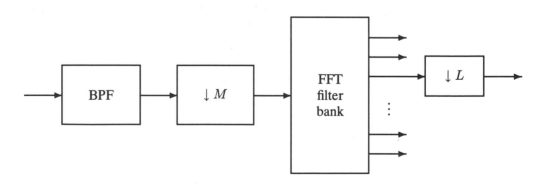

Figure 7.5. Sliding-window frequency-analysis system.

- The choice of some filter design parameters is flexible, e.g., transition widths and deviations. You should use rational judgments for these and explain your reasoning.

- Since the gain of any one channel can be regarded as a frequency sample of the DFT, we would like to guarantee that the "gain" from the input, through the BPF, and then to each DFT channel output will be one at the center frequency of the channel. This is the only constraint on the passband of the channel BPFs. As part of the channel BPF filter design, you must determine the gain compensation for each channel of the filter bank. Is the gain the same across all the output channels of the sliding-window DFT filter bank?

Part 3

(f) Evaluate the number of operations (multiplications and additions) per second for your entire system. State the output sampling rate for your filter bank analyzer.

(g) Make a tradeoff between the rate change factor following the BPF and the FFT length. Try to reduce the number of multiplications and additions required. It is not necessary (and it is probably impossible) to write an analytic formula that can be minimized, so try some different options and pick one that seems good.

7.3 REFERENCES

[1] R. W. Gerchberg, "Super-resolution through Error Energy Reduction," *Optica Acta*, Vol. 14, No. 9, pp. 709–720, Sept. 1979.

[2] A. Papoulis, "A New Algorithm in Spectral Analysis and Bandlimited Extrapolation," *IEEE Transactions on Circuits and Systems*, Vol. CAS-22, No. 9, Sept. 1975.

[3] J. L. Yen, "On Nonuniform Sampling of Bandwidth-Limited Signals," *IRE Transactions on Circuit Theory,* pp. 251–257, Dec. 1956.

[4] J. J. Clark, M. R. Palmer, and P. D. Lawrence, "A Transformation Method for the Reconstruction of Functions from Nonuniformly Spaced Samples," *IEEE Transactions on Acoustics, Speech, and Signal Processing*, Vol. ASSP-33, No. 4, pp. 1151–1165, Oct. 1985.

[5] P. Duhamel and H. Hollmann, "A 'Split-Radix' Fast Fourier Transform," *Electronics Letters*, Vol. 20, No. 1, pp. 14–16, Jan. 5, 1984.

Index